10 Minnesota MCA Grade 8 Math Practice Tests

The Ultimate Test Prep Collection with Answer Explanations

Dr. A. Nazari

Published by View Math Education

ViewMath.com

10 Practice Tests

Grade 8 Mathematics

Welcome!

*This book contains **10 full-length practice tests** — the most comprehensive preparation you can get for your Grade 8 math assessment. Each test covers all six topics:*

- *Irrational Numbers*
- *Powers & Scientific Notation*
- *Linear Equations*
- *Functions*
- *Geometry*
- *Data & Relationships*

Ten tests give you the practice needed to walk into the real test feeling fully prepared.

Thorough preparation leads to outstanding results.

> *“Ten full tests! By the time you finish, there won't be any surprises on test day.”*

How to Use This Book

A complete 10-test preparation program

What's Inside

- ***10 Full-Length Practice Tests*** — *each covers all 6 chapters of Grade 8 math: irrational numbers, exponents & scientific notation, linear equations, functions, geometry, and data analysis.*
- ***Detailed Answer Explanations*** — *every question includes a step-by-step solution so you learn from every mistake.*
- ***Formula Reference Sheet*** — *all the key Grade 8 formulas you need, organized and ready for quick review.*
- ***Test Tracker*** — *log your scores across all 10 tests and monitor your progress from start to finish.*

Your 10-Test Training Plan

★ PHASE 1: Foundation (Tests 1–3)

Untimed or soft-timed. Focus on understanding the format, identifying strengths and weaknesses, and building good study habits.

★★ PHASE 2: Building Skills (Tests 4–7)

Timed (70 minutes each). Work on pacing, accuracy, and showing complete solutions. Review weak topics between tests.

★★★ PHASE 3: Test-Day Ready (Tests 8–10)

Full test conditions: strict timing, quiet space, no notes. Compare scores with your early tests to see your growth.

Schedule: *Take one test every 3–4 days, or one per week. Use study days between tests to review.*

Types of Questions

Multiple Choice: *Four options — work the problem first, then match. Eliminate obviously wrong answers to narrow your choices.*

Short Answer & Constructed Response: *Show every step: equations, substitutions, simplifications. Partial credit rewards correct reasoning even if the final answer is off.*

Graphing & Data Analysis: *Plot points, draw lines, interpret graphs. Label axes clearly.*

Tip: Ten tests is a full preparation program. Don't rush. The key is what you do between tests — study, review, and understand your mistakes before moving forward.

Test-Taking Tips

Your complete test-day toolkit

Before the Test

- *Review your notes from the previous test — focus on your weak topics*
- *Set up a quiet, clean workspace with all your materials ready*
- *Start with a positive mindset: you've prepared for this*

During the Test

- *Read each problem fully before calculating anything*
- *Write the formula or set up the equation first, then substitute values*
- *Show all your work — every step, every operation*
- *If stuck for more than 2 minutes, mark it and move on*
- *Use estimation to check if your answers are reasonable*

After the Test

- *Read the full explanation for every question you got wrong*
- *Write down which topics gave you trouble (not just question numbers)*
- *Study those topics before taking the next test*
- *Record your score in the Test Tracker*

Common Mistakes in Grade 8 Math

Exponents: $(ab)^n = a^n b^n$, *but* $a^m + a^n \neq a^{m+n}$. *Only multiply/divide to combine.*

Slope formula: $m = \frac{y_2 - y_1}{x_2 - x_1}$ *— keep the order consistent.*

Systems of equations: *The solution must satisfy both equations.*

Transformations: *Rotations and reflections change position; dilations change size.*

Volume: *Use* $\pi \approx 3.14$ *or leave as* π *— match what the question asks.*

“The students who improve the most aren't the ones who take the most tests — they're the ones who carefully review every mistake. Make that your priority.”

What You'll Need

Gather these materials before you begin

Materials Checklist

Sharpened Pencils — #2 pencils, at least two

Good Eraser — for clean corrections

Scratch Paper — for working out problems

Ruler / Straightedge — for graphing & geometry

Quiet Space — no distractions

Focused Mind — ready to do your best

Allowed Materials

- ✓ Pencils and eraser
- ✓ Scratch paper (provided on official test day)
- ✓ Ruler or straightedge (if required)
- ✓ Protractor (if required)

Not Allowed

- ✗ Calculator (unless your state test allows it)
- ✗ Cell phone or any electronic device
- ✗ Notes, textbooks, or reference sheets
- ✗ Help from others during the test

A Note for Parents & Guardians

Ten tests is a comprehensive program. Plan ***one test every 3–4 days*** *(or one per week) with study sessions between each test.*

How to help:

- *Tests 1–3 should be untimed — build understanding before adding pressure.*
- *After each test, review the answer explanations together. Ask: "Which topics were hardest? Let's study those before the next one."*
- *Use the Test Tracker to celebrate progress over time.*
- *For topic-specific help, pair this book with our* ***Grade 8 Math Study Guide*** *or* ***Grade 8 Workbook****.*

Grade 8 Formula Reference

Keep this page handy — you may use it during your practice tests!

X¹ Exponent Rules

$a^m \cdot a^n = a^{m+n}$ $(a^m)^n = a^{mn}$ $(ab)^n = a^n \cdot b^n$

$\frac{a^m}{a^n} = a^{m-n}$ $a^0 = 1 \; (a \neq 0)$ $a^{-n} = \frac{1}{a^n}$

Lines & Linear Equations

Slope: $m = \frac{y_2 - y_1}{x_2 - x_1} = \frac{\text{rise}}{\text{run}}$

m = slope b = y-intercept

Slope-intercept: $y = mx + b$

Parallel lines: same slope

Proportional: $y = mx$

Proportional: passes through origin

Scientific Notation

$a \times 10^n$ where $1 \leq |a| < 10$

Multiply: add exponents

Divide: subtract exponents

√x Roots & Number Sense

Perfect squares: 1, 4, 9, 16, 25, 36, 49, 64, 81, 100, 121, 144

Perfect cubes: 1, 8, 27, 64, 125

$\sqrt{2} \approx 1.414$ $\sqrt{3} \approx 1.732$ $\pi \approx 3.14159$

Pythagorean Theorem & Distance

$a^2 + b^2 = c^2$ c = hypotenuse (longest side of a right triangle)

Distance: $d = \sqrt{(x_2 - x_1)^2 + (y_2 - y_1)^2}$

Volume Formulas

Cylinder $V = \pi r^2 h$ **Cone** $V = \frac{1}{3}\pi r^2 h$ **Sphere** $V = \frac{4}{3}\pi r^3$

×

Angle Relationships

Triangle angle sum: $180°$ **Exterior angle** = sum of two remote interior angles

Parallel lines + transversal: Alternate interior angles are equal • Co-interior angles sum to $180°$

Functions

Each input → exactly one output **Vertical line test:** if any vertical line hits graph more than once ⇒ not a function

Linear: constant rate of change ($y = mx + b$) **Nonlinear:** rate of change varies

Transformations

Translation: slide **Reflection:** flip **Rotation:** turn **Dilation:** resize

Congruent = same shape & size Similar = same shape, proportional size

Tip: Bookmark this page! Review it before each test so these formulas become second nature.

Multiplication Table

You may use this table during your practice tests!

×	1	2	3	4	5	6	7	8	9	10	11	12
1	1	2	3	4	5	6	7	8	9	10	11	12
2	2	4	6	8	10	12	14	16	18	20	22	24
3	3	6	9	12	15	18	21	24	27	30	33	36
4	4	8	12	16	20	24	28	32	36	40	44	48
5	5	10	15	20	25	30	35	40	45	50	55	60
6	6	12	18	24	30	36	42	48	54	60	66	72
7	7	14	21	28	35	42	49	56	63	70	77	84
8	8	16	24	32	40	48	56	64	72	80	88	96
9	9	18	27	36	45	54	63	72	81	90	99	108
10	10	20	30	40	50	60	70	80	90	100	110	120
11	11	22	33	44	55	66	77	88	99	110	121	132
12	12	24	36	48	60	72	84	96	108	120	132	144

How to Use This Table

To find **4** × **7**:

1. Find **4** in the left column (blue).
2. Find **7** in the top row (blue).
3. Follow the row and column until they meet: the answer is **28**!

Tip: You can also use this table for division! If you know $28 \div 4 = ?$, find 28 in the 4's row. The column header gives you the answer: **7**!

My Test Tracker

Record every test and watch your scores improve

Name: ______________________ **Start Date:** ______________

GETTING STARTED (Tests 1–3)

Test 1 — Untimed

Date: __________ Score: ______/______ %: ______ Topics to review: ______________

Test 2 — Untimed

Date: __________ Score: ______/______ %: ______ Topics to review: ______________

Test 3 — Soft Timer

Date: __________ Score: ______/______ %: ______ Topics to review: ______________

BUILDING SKILLS (Tests 4–7)

Test 4 — Timed (70 min)

Date: __________ Score: ______/______ %: ______ Focus area: ______________

Test 5 — Timed (70 min)

Date: __________ Score: ______/______ %: ______ Focus area: ______________

Test 6 — Timed (70 min)

Date: __________ Score: ______/______ %: ______ Focus area: ______________

Test 7 — Timed (70 min)

Date: __________ Score: ______/______ %: ______ Focus area: ______________

TEST-DAY READY (Tests 8–10)

Test 8 — Full Test Conditions

Date: ____________ Score: _______/_______ %: _______ Growth since Test 1: ____________________

Test 9 — Full Test Conditions

Date: ____________ Score: _______/_______ %: _______ Growth since Test 1: ____________________

Test 10 — Full Test Conditions

Date: ____________ Score: _______/_______ %: _______ Growth since Test 1: ____________________

Score Progress

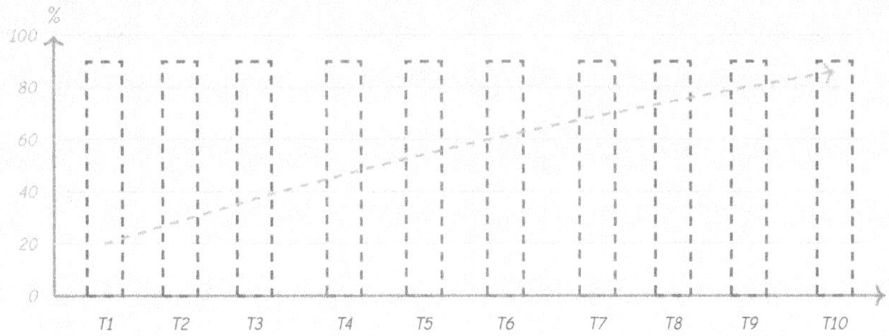

Shade each bar after every test. Watch your improvement!

Final Reflection

The most important thing I learned: ____

The topic where I improved the most: ____

My advice for other students: ____

YOUR ONLINE COMPANION

Continue Learning at ViewMath Academy!

For Parents, Teachers & Students

Great job on the practice tests! Want to keep improving? ViewMath Academy is your ***free online companion*** *to this book.*

- ***Score Analyzer*** — *Enter your answers and instantly see which topics need more practice*
- ***Interactive Lessons*** — *Review the concepts behind each question with clear explanations*
- ***Adaptive Quizzes*** — *Practice your weak topics with questions that match your level*
- ***Progress Tracking*** — *See your mastery grow across all Grade 8 math topics*
- ***Personalized Dashboard*** — *A learning plan tailored just for you*

Scan to visit ViewMath Academy

ViewMath.com/MN-Grade8

Free to use • No downloads required • Works on any device

★ Table of Contents ★

Here's what we'll explore together!

Let's learn and have fun!

PRACTICE TEST

1

Practice Test 1

30 Questions

Before You Start

- **Read each question carefully** before choosing your answer.
- **Show your work** on scratch paper when you need to.
- **Skip hard questions** and come back to them later.
- **Check your answers** when you're done.
- **Take your time** — there's no rush!

You've Got This!

Do your best and show what you know!

1. True or false: The number $\frac{22}{7}$ is equal to π.

Your Answer

2. The table below shows four repeating decimals and the power of 10 a student used to convert each. Which one uses the WRONG power of 10?

Decimal	Multiply by
$0.\overline{7}$	10
$0.\overline{45}$	100
$0.\overline{123}$	100
$0.\overline{8}$	10

(A) $0.\overline{7}$

(B) $0.\overline{45}$

(C) $0.\overline{123}$

(D) $0.\overline{8}$

3. The table below shows a student's work to approximate $\sqrt{14}$. What should go in the blank?

Guess	Square	Compare to 14
3	9	too small
4	16	too big
3.7	13.69	too small
3.8	14.44	too big
?	?	closest

(A) 3.74, because $3.74^2 = 13.9876$

(B) 3.72, because $3.72^2 = 13.8384$

(C) 3.80, because $3.8^2 = 14.44$

(D) 3.75, because $3.75^2 = 14.0625$

4. *What is the best estimate for* $5 - \sqrt{7}$*?*

(A) 1.4 (B) 2.4

(C) 3.0 (D) 0.4

5. *The diagram shows a square and a cube. The square has an area of* A *square units and the cube has a volume of* V *cubic units. If* $A = V$*, which pair of side lengths is correct?*

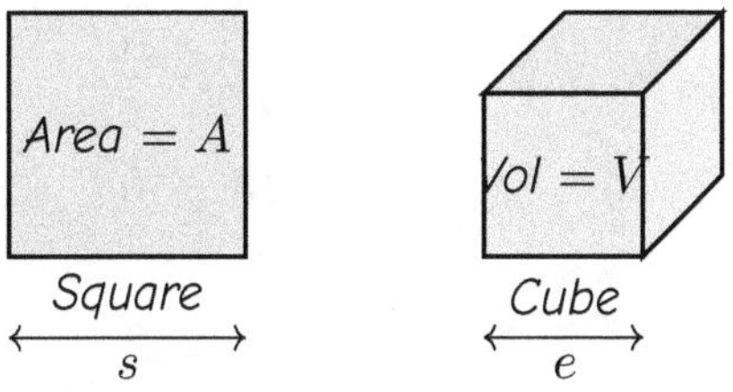

(A) $s = 6,\ e = 6$ (B) $s = 8,\ e = 4$

(C) $s = 9,\ e = 3$ (D) $s = 5,\ e = 5$

6. *Estimate the product* $498{,}000{,}000 \times 6{,}200$ *by first writing each factor in scientific notation and then computing.*

Your Answer:

7. *Study the diagram below. Each box produces its output by multiplying the two inputs. What belongs in the output box marked "?"?*

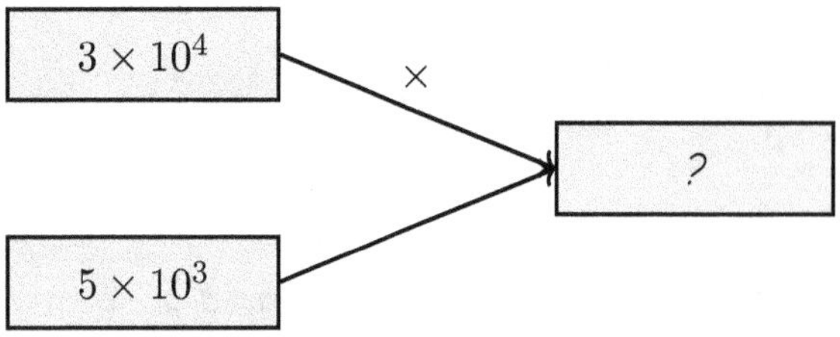

(A) 8×10^7

(B) 1.5×10^7

(C) 1.5×10^8

(D) 15×10^7

8. *A bike rental company charges \$8 per hour. Which equation models the total cost y for x hours?*

(A) $y = x + 8$

(B) $y = 8x$

(C) $y = \frac{x}{8}$

(D) $y = 8x + 10$

9. *A gym charges a \$25 joining fee plus \$10 per month. Write an equation for the total cost y after x months.*

Your Answer:

10. *Solve: $x + y = 15$ and $x - y = 3$.*

Your Answer:

11. Use the graph of g below. For which input is $g(x) = 0$?

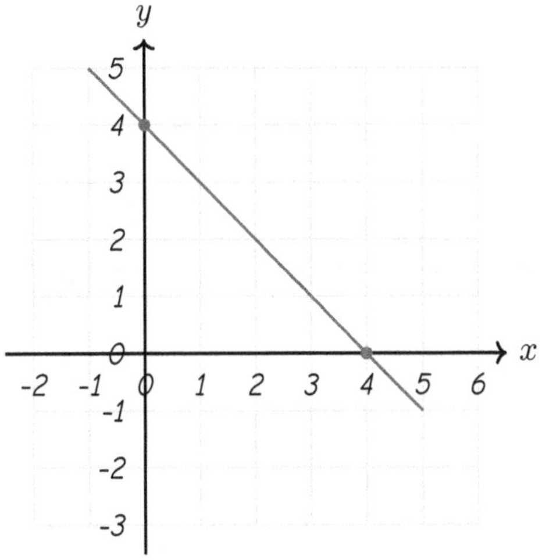

(A) $x = 0$

(B) $x = 2$

(C) $x = 4$

(D) $x = -4$

12. The tables below show two functions. Find the rate of change and initial value of each. Which function will have a greater value at $x = 10$?

Function A

x	y
0	20
1	23
2	26
3	29

Function B

x	y
0	5
1	10
2	15
3	20

Your Answer:

13. Does the equation $y = -12$ represent a linear function? Write Yes or No.

Your Answer:

14. *A taxi charges a $\$4$ flat fee plus $\$3$ per mile. Which equation models the total cost y for x miles?*

(A) $y = 4x + 3$

(B) $y = 3x + 4$

(C) $y = 7x$

(D) $y = 3x - 4$

15. *A graph shows a line going upward with a constant slope. Is the rate of change constant or varying?*

Your Answer:

16. *Which of the following is NOT preserved under a translation?*

(A) *Side lengths*

(B) *Angle measures*

(C) *Position*

(D) *Parallelism of sides*

17. *Describe a sequence of rigid transformations that maps a triangle with vertices $(1, 1)$, $(4, 1)$, $(1, 3)$ to a triangle with vertices $(-1, 1)$, $(-4, 1)$, $(-1, 3)$.*

Your Answer:

18. *Point $(5, 2)$ is rotated $180°$ around the origin. What are the coordinates of the image?*

Your Answer:

19. *A dilation centered at the origin maps $(4, -2)$ to $(12, -6)$. What is the scale factor?*

(A) 2

(B) 3

(C) 4

(D) 6

20. Parallel lines ℓ and m are cut by transversal t. If a pair of alternate interior angles are $(4x)^\circ$ and 80°, what is x?

(A) 320
(B) 20
(C) 25
(D) 10

21. A right triangle has legs 5 and 5. What is the hypotenuse?

(A) 10
(B) $\sqrt{50}$
(C) 25
(D) $\sqrt{10}$

22. What is the distance between $(2, -5)$ and $(6, -2)$?

(A) 25
(B) 7
(C) 5
(D) $\sqrt{7}$

23. A cylinder has a volume of 450π cm^3. A cone has the same base and height. What is the cone's volume?

(A) 150π cm^3
(B) 225π cm^3
(C) 1350π cm^3
(D) 900π cm^3

24. How many square pyramids with base side 4 cm and height 6 cm could fill a rectangular prism with dimensions $4 \times 4 \times 6$ cm?

(A) 1
(B) 2
(C) 3
(D) 4

25. As temperature increases, heating costs decrease. This relationship is:

(A) a positive linear association

(B) a negative linear association

(C) no association

(D) a nonlinear association

26. A scatter plot shows these points with a trend line through $(1, 3)$ and $(5, 7)$. What is the equation of the trend line?

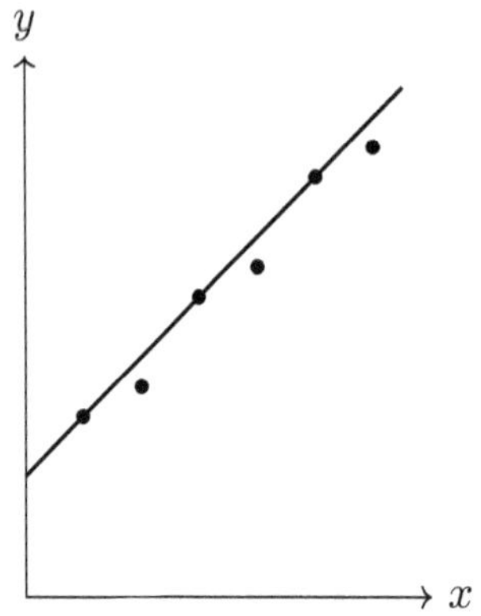

(A) $y = x + 2$

(B) $y = 2x + 1$

(C) $y = x + 3$

(D) $y = 0.5x + 2.5$

27. Why can extrapolation be unreliable?

(A) Because you cannot use a calculator

(B) Because the pattern may not continue outside the observed data range

(C) Because the slope is always wrong

(D) Because the y-intercept is always zero

28. Using the table from q19, what percentage of boys prefer reading? What percentage of girls prefer reading? Is there an association?

Your Answer

29. $P(\text{event}) = \frac{3}{4}$. What is $P(\text{not event})$?

(A) $\frac{3}{4}$

(B) $\frac{1}{4}$

(C) $\frac{1}{2}$

(D) 0

30. Explain why $\binom{5}{2} = \binom{5}{3}$.

Your Answer:

★ End of Practice Test 1 ★

Great job finishing the test!

My Score

I got __________ out of 30 questions right.

*Check your answers in the **Answer Key** at the back of the book.*

Review any questions you missed. That's how we learn!

Check Your Score Online!

*Visit **ViewMath Academy** to enter your answers and see which topics you need to review. You can also explore lessons, take quizzes, track your scores, and save your progress!*

viewmath.com/score/8.1.MN.16

*Or go to **viewmath.com/score** and enter code: **8.1.MN.16***

PRACTICE TEST

2

Practice Test 2

30 Questions

Before You Start

- **Read each question carefully** before choosing your answer.
- **Show your work** on scratch paper when you need to.
- **Skip hard questions** and come back to them later.
- **Check your answers** when you're done.
- **Take your time** — there's no rush!

You've Got This!

Do your best and show what you know!

1. *The number line below shows two points, P and Q.*

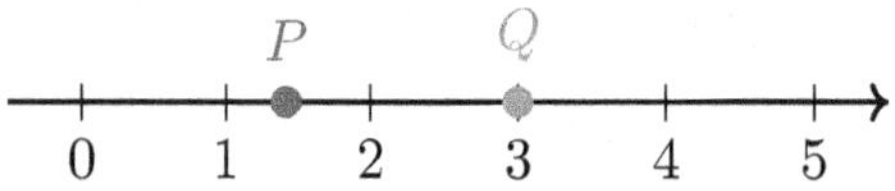

Point P represents $\sqrt{2}$ and point Q represents 3. Which statement is true?

(A) Both P and Q are rational.

(B) Both P and Q are irrational.

(C) P is irrational and Q is rational.

(D) P is rational and Q is irrational.

2. *The flowchart below shows the conversion steps for a repeating decimal. Fill in the missing values to convert $0.\overline{54}$ to a fraction.*

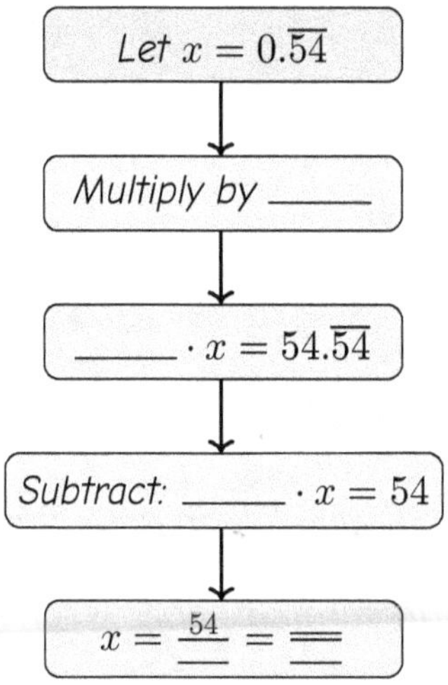

Your Answer:

3. *The squares below have the given areas. Estimate the side length of each square to one decimal place.*

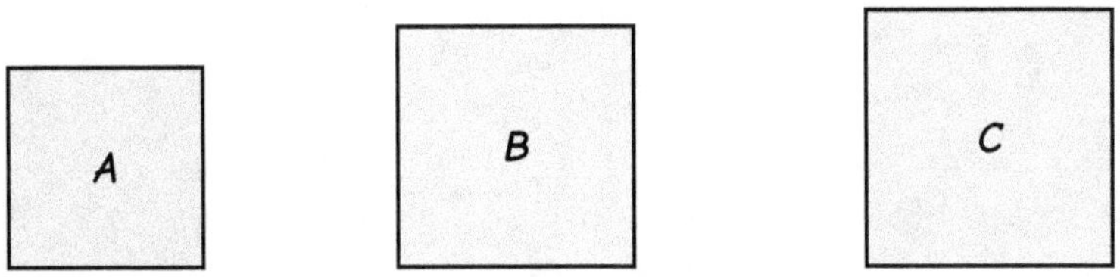

Area $= 18$ *sq units* *Area* $= 40$ *sq units* *Area* $= 72$ *sq units*

Your Answer:

4. *What is the best estimate for* $\sqrt{5} + \sqrt{3}$?

(A) $\sqrt{8} \approx 2.8$

(B) 3.0

(C) 4.0

(D) 8.0

5. *Solve* $x^2 = \frac{9}{16}$.

(A) $x = \frac{3}{4}$ *only*

(B) $x = \pm\frac{3}{4}$

(C) $x = \frac{9}{8}$

(D) $x = \pm\frac{9}{4}$

6. *Which of the following is true about* 4.2×10^{-3} *and* 4.2×10^{3}?

(A) *They are equal*

(B) 4.2×10^{3} *is* 10^6 *times as large as* 4.2×10^{-3}

(C) 4.2×10^{-3} *is* 10^6 *times as large as* 4.2×10^{3}

(D) *Their product is* 1

7. *What is* $\frac{8\times10^7}{2\times10^3}$?

(A) 4×10^4

(B) 6×10^4

(C) 4×10^{10}

(D) 16×10^{10}

8. A faucet leaks at a constant rate. It leaks 45 mL in 15 minutes. How many mL does it leak in 1 hour?

Your Answer:

9. A line passes through $(1,2)$ and $(4,11)$. What is the equation in slope-intercept form?

Ⓐ $y = 3x - 1$

Ⓑ $y = 3x + 1$

Ⓒ $y = 3x + 2$

Ⓓ $y = 3x - 2$

10. Solve: $3x + y = 10$ and $y = x - 2$.

Your Answer:

11. The function $f(x) = 10 - 3x$ gives the number of pieces of candy remaining after x friends each take 3. How many pieces are left after 2 friends take candy?

Ⓐ 1

Ⓑ 4

Ⓒ 7

Ⓓ 16

12. A table for Function C shows $(0,7)$, $(1,11)$, $(2,15)$. What is the initial value?

Your Answer:

13. Look at the graph below. Is the function linear or nonlinear? Explain your reasoning.

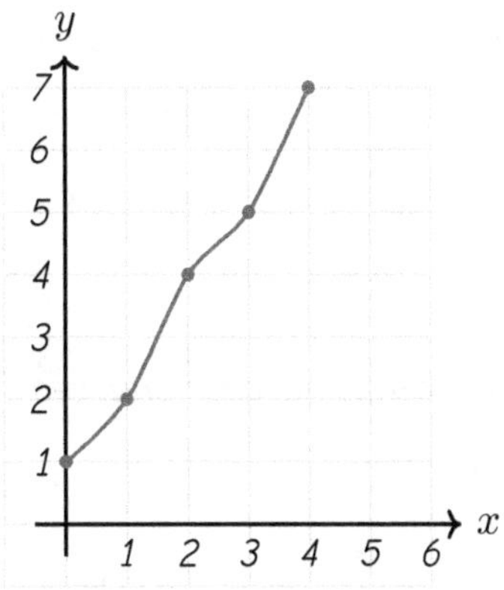

Your Answer:

14. The table shows $(2, 9)$ and $(6, 25)$. Find the slope of the line.

Your Answer:

15. A graph is a straight line segment going upward. The function during this segment is:

(A) Nonlinear and increasing

(B) Linear and decreasing

(C) Linear and increasing

(D) Nonlinear and constant

16. A figure is reflected over the y-axis. What happens to point $(7, -2)$?

(A) $(7, 2)$

(B) $(-7, 2)$

(C) $(-7, -2)$

(D) $(-2, 7)$

17. Triangle P has angles $40°$, $60°$, and $80°$. Triangle Q has angles $40°$, $60°$, and $80°$. Are they congruent?

(A) Yes, because matching angles guarantee congruence.

(B) No, because angles alone do not guarantee congruence.

(C) Yes, because all three angles match.

(D) No, because the angles are in different order.

18. A point is translated by $(3, -2)$ and then reflected over the y-axis. If the original point is $(1, 4)$, what is the final image?

(A) $(-4, 2)$

(B) $(4, 2)$

(C) $(-4, -2)$

(D) $(4, -2)$

19. Two similar rectangles have a scale factor of 3. The smaller has area 12 cm^2. What is the area of the larger?

Your Answer:

20. Two parallel lines are cut by a transversal. One angle measures $72°$. What is its alternate interior angle?

(A) $18°$

(B) $72°$

(C) $108°$

(D) $288°$

21. A right triangle has legs 8 and 15. What is the hypotenuse?

(A) 23

(B) 17

(C) $\sqrt{23}$

(D) 289

22. Find the distance between $(5, -2)$ and $(5, 7)$.

Your Answer:

23. What is the volume of a cylinder with radius 7 m and height 3 m? Leave your answer in terms of π.

(A) 21π m^3

(B) 63π m^3

(C) 147π m^3

(D) 49π m^3

24. A triangular pyramid has a base with legs 5 cm and 12 cm (a right triangle) and a height of 10 cm. Find the volume.

Your Answer:

25. Identify the outlier in the scatter plot.

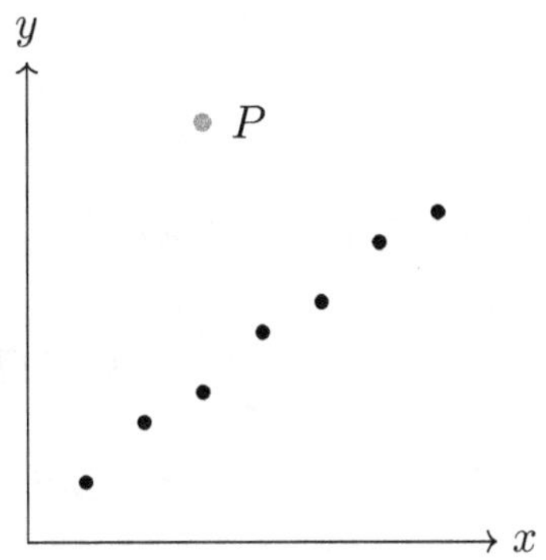

(A) $(1, 1)$

(B) $(7, 5.5)$

(C) $(3, 7)$ — point P

(D) $(4, 3.5)$

26. Draw a line of best fit through the data and estimate its slope.

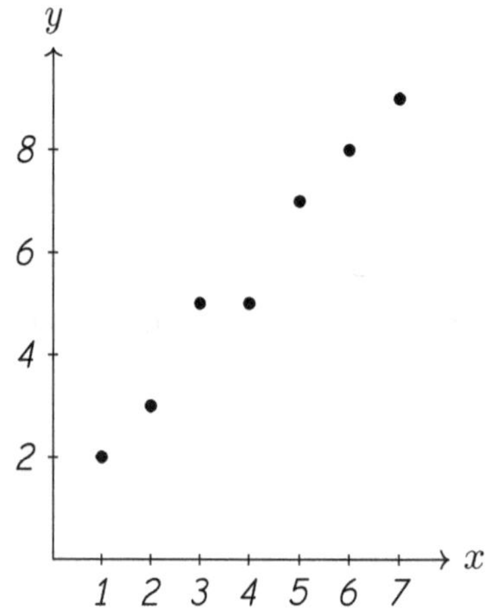

Your Answer:

27. A linear model is $y = 2x + 5$. What does the slope represent?

(A) The starting value

(B) The rate of change per unit increase in x

(C) The x-intercept

(D) The total value of y

28. A two way table has 3 rows and 4 columns (not counting totals). How many joint frequencies are there?

(A) 3

(B) 4

(C) 7

(D) 12

29. A tree diagram shows flipping a coin and rolling a die (1–6). How many total outcomes are in the sample space?

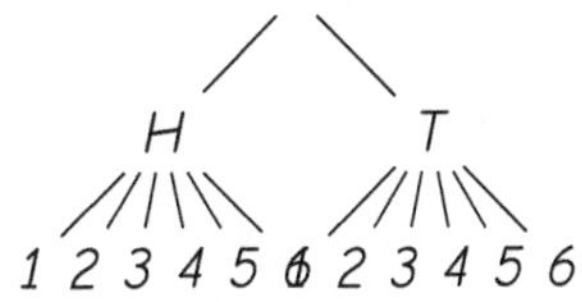

(A) 6

(B) 8

(C) 12

(D) 36

30. The diagram shows all arrangements of the letters A, B, C. How many permutations are there?

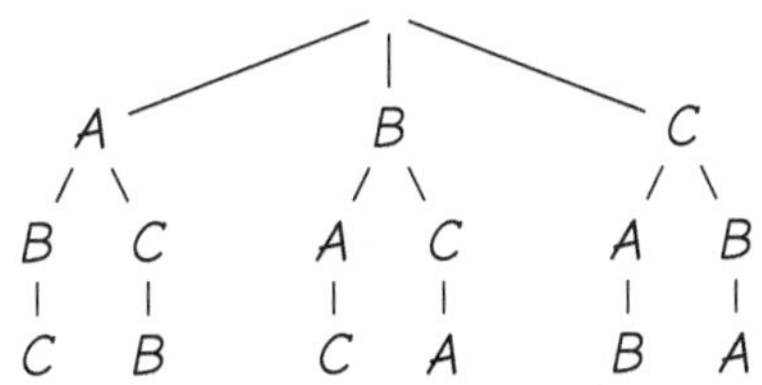

(A) 3

(B) 6

(C) 9

(D) 12

End of Practice Test 2

Great job finishing the test!

My Score

I got __________ out of 30 questions right.

*Check your answers in the **Answer Key** at the back of the book.*

Review any questions you missed. That's how we learn!

Check Your Score Online!

*Visit **ViewMath Academy** to enter your answers and see which topics you need to review. You can also explore lessons, take quizzes, track your scores, and save your progress!*

viewmath.com/score/8.1.MN.17

*Or go to **viewmath.com/score** and enter code: **8.1.MN.17***

PRACTICE TEST

3

Practice Test 3

30 Questions

Before You Start

- **Read each question carefully** before choosing your answer.
- **Show your work** on scratch paper when you need to.
- **Skip hard questions** and come back to them later.
- **Check your answers** when you're done.
- **Take your time** — there's no rush!

You've Got This!

Do your best and show what you know!

1. Which number below can be written as a fraction $\frac{a}{b}$ where a and b are integers and $b \neq 0$?

(A) $\sqrt{3}$

(B) π

(C) $0.\overline{81}$

(D) $\sqrt{11}$

2. What is $0.\overline{27}$ as a fraction in simplest form?

(A) $\frac{27}{100}$

(B) $\frac{27}{99}$

(C) $\frac{3}{11}$

(D) $\frac{9}{33}$

3. A student estimated $\sqrt{60} \approx 8$. Is this estimate reasonable?

(A) Yes, because $8^2 = 64$ is close to 60.

(B) No, because $8^2 = 64$, which is too far from 60. A better estimate is about 7.7.

(C) No, $\sqrt{60} = 30$.

(D) Yes, because $60 \div 8 \approx 8$.

4. Which is larger: $4\sqrt{2}$ or $\sqrt{30}$?

(A) $4\sqrt{2}$, because $4 > \sqrt{30}$

(B) $\sqrt{30}$, because $30 > 2$

(C) $4\sqrt{2}$, because $4\sqrt{2} \approx 5.66 > 5.48 \approx \sqrt{30}$

(D) They are equal.

5. A square has an area of 196 square centimeters. What is the length of one side?

(A) 13 cm

(B) 49 cm

(C) 14 cm

(D) 98 cm

6. Which of the following is 0.00072 written in scientific notation?

(A) 7.2×10^{-4}

(B) 7.2×10^{4}

(C) 72×10^{-5}

(D) 0.72×10^{-3}

7. A virus has a mass of 9.5×10^{-18} grams. If a sample contains 2×10^{6} viruses, what is the total mass?

(A) 1.9×10^{-11} g

(B) 1.9×10^{-12} g

(C) 11.5×10^{-12} g

(D) 1.9×10^{-24} g

8. Store A sells apples for $2 per pound. Store B's prices are shown in the table.

Pounds	3	6	9
Cost ($)	7.50	15.00	22.50

Which store has the lower unit price?

(A) Store A

(B) Store B

(C) They charge the same price.

(D) Not enough information.

9. Which equation has the steepest line?

(A) $y = x + 1$

(B) $y = 3x - 2$

(C) $y = -5x + 4$

(D) $y = \frac{1}{2}x + 6$

10. How many solutions does this system have? $y = 2x + 5$ and $4x - 2y = -10$.

(A) No solution

(B) One solution

(C) Two solutions

(D) Infinitely many solutions

11. *A function is defined by the table below.*

x	0	1	2	3
$f(x)$	-1	3	7	11

What is $f(0) + f(2)$?

(A) 2
(B) 6
(C) 8
(D) 10

12. *Function A: $y = -x + 10$. Function B: $y = -4x + 10$. Which function decreases faster? Write A or B.*

Your Answer:

13. *A table shows $(0, 3)$, $(1, 7)$, $(2, 11)$, $(3, 15)$. Is this linear or nonlinear?*

(A) *Nonlinear, because the outputs are odd numbers only.*
(B) *Linear, because the change in y is always 4.*
(C) *Nonlinear, because the outputs get larger.*
(D) *Linear, because the first output is positive.*

14. *A line passes through $(0, 4)$ and $(5, 24)$. Write its equation.*

Your Answer:

15. *A graph of water in a tank is a straight line sloping downward. What is happening?*

(A) *Water is being added at a constant rate.*
(B) *Water is leaking out at a constant rate.*
(C) *The tank is full and not changing.*
(D) *Water is being added at an increasing rate.*

16. Point $R(-1,4)$ is rotated $90°$ counterclockwise around the origin. What are the coordinates of R'?

(A) $(4,1)$

(B) $(-4,-1)$

(C) $(1,-4)$

(D) $(-4,1)$

17. Two quadrilaterals are shown below. Which statement is correct?

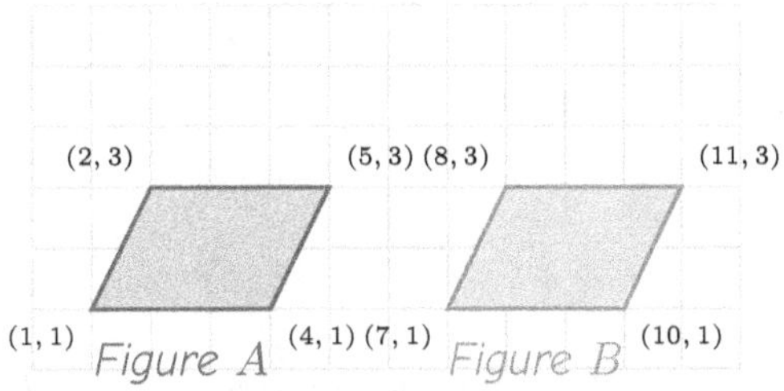

(A) They are congruent — mapped by a translation of 6 units right.

(B) They are not congruent — different side lengths.

(C) They are similar but not congruent.

(D) They are congruent — mapped by a reflection.

18. Point $A(0,-5)$ is rotated $90°$ counterclockwise around the origin. Where does A' land?

(A) $(5,0)$

(B) $(0,5)$

(C) $(-5,0)$

(D) $(0,-5)$

19. Two similar triangles have a scale factor of 4 from the smaller to the larger. If a side of the smaller triangle is 3 cm, what is the corresponding side of the larger triangle?

(A) 7 cm

(B) 0.75 cm

(C) 12 cm

(D) 9 cm

20. *In an isosceles triangle, the two base angles are each $54°$. What is the vertex angle?*

(A) $54°$

(B) $72°$

(C) $108°$

(D) $126°$

21. *A baseball diamond is a square with 90 ft sides. How far is it from home plate to second base (the diagonal)?*

(A) 180 *ft*

(B) $90\sqrt{2}$ *ft*

(C) 45 *ft*

(D) $\sqrt{90}$ *ft*

22. *Find the distance between points A and B shown on the coordinate plane.*

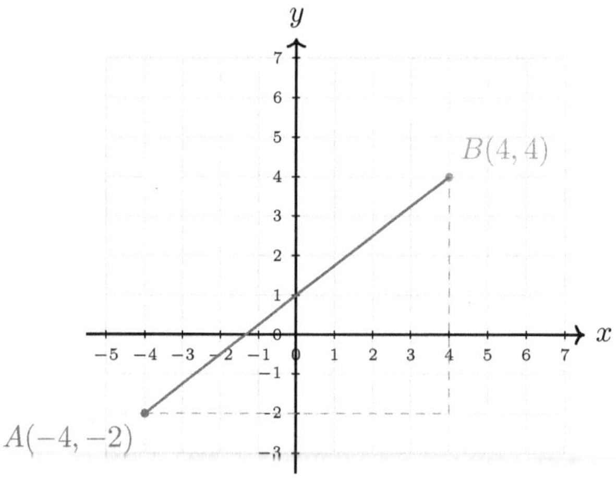

Your Answer:

23. *The volume of a sphere with radius r is $\frac{4}{3}\pi r^3$. If the radius is tripled, the new volume is:*

(A) 3 *times the original*

(B) 9 *times the original*

(C) 27 *times the original*

(D) 81 *times the original*

24. A square pyramid has base side 7 cm and height 12 cm. What is the volume?

(A) $196\ cm^3$
(B) $588\ cm^3$
(C) $294\ cm^3$
(D) $98\ cm^3$

25. A scatter plot has a strong upward trend. Which r-value (correlation) is most likely?

(A) $r = -0.9$
(B) $r = 0.1$
(C) $r = 0.9$
(D) $r = 0$

26. Data: $(1, 2), (2, 4), (3, 7), (4, 8), (5, 11)$. Estimate the slope of a reasonable trend line.

Your Answer:

27. A positive residual means:

(A) The predicted value is higher than the actual value.
(B) The actual value is higher than the predicted value.
(C) The slope of the model is positive.
(D) The data point is an outlier.

28. A survey asks students about their favorite sport (soccer or basketball) and grade (7th or 8th). 20 7th-graders chose soccer. This 20 is a:

(A) marginal frequency
(B) joint frequency
(C) relative frequency
(D) total frequency

29. Two events are independent if:

(A) They cannot happen at the same time
(B) The outcome of one does not affect the other
(C) They always happen together
(D) They share the same sample space

30. How many 3-letter codes can be made from the letters A, B, C, D if no letter is repeated?

(A) 12

(B) 24

(C) 64

(D) 4

End of Practice Test 3

Great job finishing the test!

My Score

I got __________ out of 30 questions right.

*Check your answers in the **Answer Key** at the back of the book.*

Review any questions you missed. That's how we learn!

Check Your Score Online!

*Visit **ViewMath Academy** to enter your answers and see which topics you need to review. You can also explore lessons, take quizzes, track your scores, and save your progress!*

viewmath.com/score/8.1.MN.18

*Or go to **viewmath.com/score** and enter code: **8.1.MN.18***

PRACTICE TEST

4

Practice Test 4

30 Questions

Before You Start

- **Read each question carefully** before choosing your answer.
- **Show your work** on scratch paper when you need to.
- **Skip hard questions** and come back to them later.
- **Check your answers** when you're done.
- **Take your time** — there's no rush!

You've Got This!

Do your best and show what you know!

1. Which of the following numbers is irrational?

(A) $\frac{5}{8}$

(B) $0.\overline{6}$

(C) $\sqrt{9}$

(D) $\sqrt{7}$

2. Write $0.5\overline{8}$ as a fraction in simplest form.

Your Answer:

3. Which inequality is true?

(A) $\sqrt{15} < 3$

(B) $\sqrt{15} > 4$

(C) $\sqrt{15} > 3$

(D) $\sqrt{15} = 4$

4. Compute $\sqrt{2} \times \sqrt{18}$ exactly, without a calculator.

Your Answer:

5. What is $\sqrt{49}$?

(A) 6

(B) 7

(C) 8

(D) 24.5

6. What is 6.1×10^3 written in standard form?

(A) 61,000

(B) 610

(C) 6,100

(D) 0.0061

7. A bacterium has a length of 2×10^{-6} m. If 1 micrometer (μm) $= 10^{-6}$ m, what is the bacterium's length in micrometers?

(A) 2×10^{-12} μm

(B) 0.002 μm

(C) 2 μm

(D) $2{,}000$ μm

8. Taxi A charges \$3 per mile. Taxi B charges \$36 for a 9-mile trip. Which taxi has the higher unit rate?

Your Answer:

9. Write the equation of a line with slope $\frac{2}{3}$ that passes through $(0, -4)$.

(A) $y = \frac{2}{3}x + 4$

(B) $y = \frac{2}{3}x - 4$

(C) $y = -4x + \frac{2}{3}$

(D) $y = 4x + \frac{2}{3}$

10. Solve: $y = -x + 5$ and $y = 2x - 1$.

(A) $(1, 4)$

(B) $(2, 3)$

(C) $(3, 2)$

(D) $(4, 1)$

11. If $h(x) = 2x^2$, what is $h(5)$?

(A) 20

(B) 25

(C) 50

(D) 100

12. Function A: $y = 2x + 15$. Function B: $y = 5x + 6$. What is the value of each function at $x = 3$?

Your Answer:

13. *The table below shows a function. Is it linear or nonlinear?*

x	0	1	2	3	4
y	1	2	5	10	17

(A) *Linear, because y always increases.*

(B) *Linear, because x increases by 1 each time.*

(C) *Nonlinear, because the differences in y are $1, 3, 5, 7$ — not constant.*

(D) *Nonlinear, because the outputs are all positive.*

14. *Find the equation of the line through $(1, 3)$ and $(4, 12)$.*

Your Answer:

15. *A bathtub fills up, then someone soaks, then the water drains. Which describes the water level graph?*

(A) *Decreasing, constant, increasing*

(B) *Increasing, decreasing, constant*

(C) *Increasing, constant, decreasing*

(D) *Constant, increasing, decreasing*

16. Look at the coordinate plane. Which transformation maps triangle PQR to triangle $P'Q'R'$?

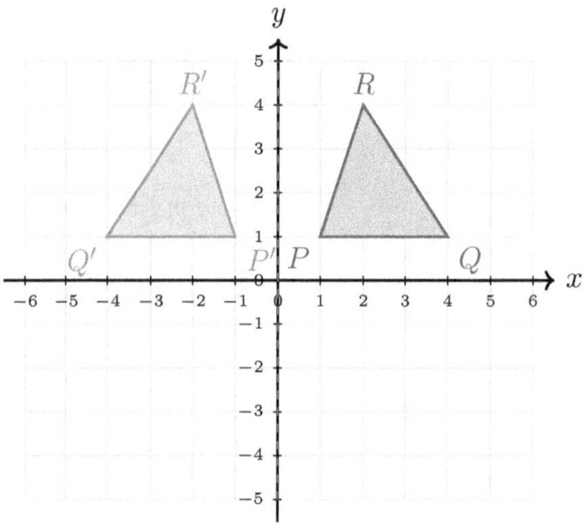

(A) Translation 5 units left

(B) Reflection over the y-axis

(C) Rotation $90°$ clockwise

(D) Reflection over the x-axis

17. Rectangle $ABCD$ has $AB = 6$ and $BC = 10$. Rectangle $WXYZ$ has $WX = 10$ and $XY = 6$. Are they congruent?

(A) No, because the side lengths are in different order.

(B) No, because $6 \neq 10$.

(C) Yes, because the sets of side lengths are the same.

(D) Cannot be determined.

18. A rectangle has vertex $(-6, 2)$. After a translation of $(4, -5)$, what is the new position of this vertex?

(A) $(-10, 7)$

(B) $(-2, -3)$

(C) $(2, -3)$

(D) $(-2, 3)$

19. *If two figures are congruent, are they also similar?*

(A) *Yes, with scale factor* $k = 1$.

(B) *No, congruent and similar are different properties.*

(C) *Only if they are triangles.*

(D) *Only if they have the same orientation.*

20. *An exterior angle of a triangle is* $140°$. *What is the adjacent interior angle?*

(A) $40°$

(B) $140°$

(C) $50°$

(D) $220°$

21. *Is a triangle with sides* 7, 24, 25 *a right triangle?*

(A) *Yes, because* $7 + 24 > 25$.

(B) *Yes, because* $7^2 + 24^2 = 25^2$.

(C) *No, because* $7^2 + 24^2 \neq 25^2$.

(D) *No, because all sides must be equal.*

22. *What is the distance between* $(-1, 2)$ *and* $(2, 6)$*?*

(A) 25

(B) 7

(C) 5

(D) $\sqrt{13}$

23. *What is the volume of the cone shown below in terms of π?*

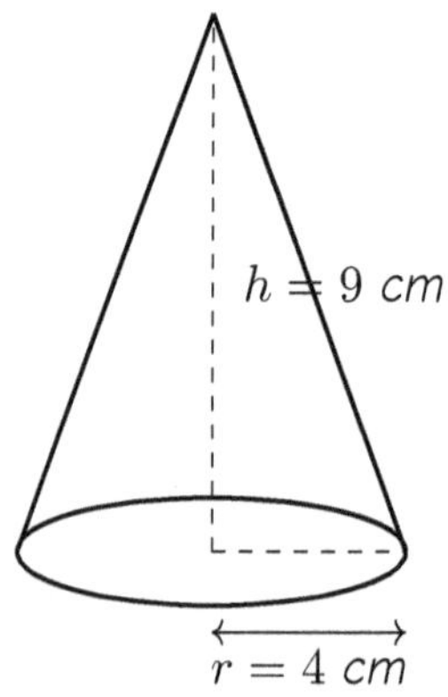

(A) $144\pi\ cm^3$

(B) $48\pi\ cm^3$

(C) $36\pi\ cm^3$

(D) $108\pi\ cm^3$

24. *A pyramid has a pentagonal base with area $35\ cm^2$ and height 6 cm. What is the volume?*

Your Answer:

25. *A scatter plot of shoe size versus favorite color shows dots scattered randomly with no pattern. What type of association is this?*

(A) Positive

(B) Negative

(C) Linear

(D) No association

26. *A trend line passes through $(3,7)$ and $(9,19)$. What is the slope?*

(A) 1

(B) 2

(C) 3

(D) 4

27. In the model $y = 2.5x + 1$, the actual value at $x = 4$ is $y = 12$. What is the residual?

(A) 0

(B) 1

(C) -1

(D) 11

28.

	Pass	Fail	Total
Studied	30	5	35
Did not study	10	15	25
Total	40	20	60

What fraction of all students passed?

(A) $\frac{30}{60}$

(B) $\frac{40}{60}$

(C) $\frac{35}{60}$

(D) $\frac{20}{60}$

29. A bag has 4 red and 6 blue marbles. Draw a tree diagram for picking 2 marbles without replacement. Find P(both blue).

Your Answer:

30. A class of 10 students needs to select 2 for a project. How many combinations?

(A) 20

(B) 45

(C) 90

(D) 100

End of Practice Test 4

Great job finishing the test!

My Score

I got __________ out of 30 questions right.

*Check your answers in the **Answer Key** at the back of the book.*

Review any questions you missed. That's how we learn!

Check Your Score Online!

*Visit **ViewMath Academy** to enter your answers and see which topics you need to review. You can also explore lessons, take quizzes, track your scores, and save your progress!*

viewmath.com/score/8.1.MN.19

*Or go to **viewmath.com/score** and enter code: **8.1.MN.19***

PRACTICE TEST

5

Practice Test 5

30 Questions

Before You Start

- **Read each question carefully** before choosing your answer.
- **Show your work** on scratch paper when you need to.
- **Skip hard questions** and come back to them later.
- **Check your answers** when you're done.
- **Take your time** — there's no rush!

You've Got This!

Do your best and show what you know!

1. A calculator shows $\sqrt{144} = 12$. Is $\sqrt{144}$ rational or irrational? Explain.

Your Answer:

2. To convert $0.\overline{63}$ to a fraction, which equation should you set up after letting $x = 0.\overline{63}$?

(A) $10x - x = 63$

(B) $100x - x = 63$

(C) $100x - x = 6.3$

(D) $1000x - x = 63$

3. Which inequality correctly describes the location of $\sqrt{22}$?

(A) $4 < \sqrt{22} < 5$

(B) $5 < \sqrt{22} < 6$

(C) $10 < \sqrt{22} < 12$

(D) $\sqrt{22} = 11$

4. The diagram below shows a rectangle with irrational side lengths. Which is the best estimate of the area?

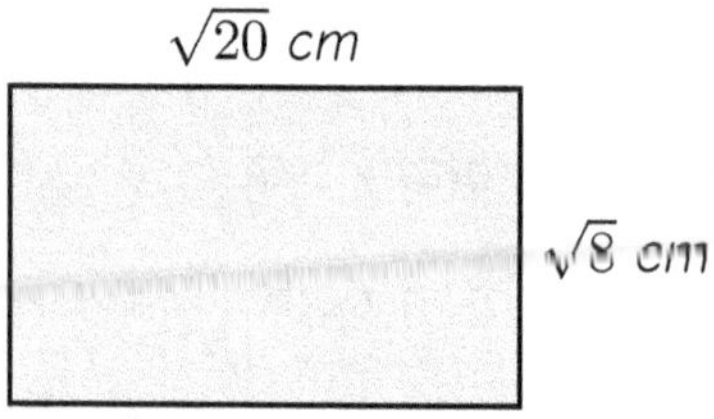

(A) $\sqrt{28} \approx 5.3$ cm^2

(B) $\sqrt{160} \approx 12.6$ cm^2

(C) 28 cm^2

(D) 160 cm^2

5. Which of the following is a perfect square?

(A) 50

(B) 72

(C) 81

(D) 90

6. *A hydrogen atom has a radius of about* 2.5×10^{-11} *meters. How many places do you move the decimal to write this in standard form?*

Your Answer:

7. *What is* $5.2 \times 10^6 + 3.8 \times 10^6$*?*

(A) 9×10^6

(B) 9×10^{12}

(C) 19.76×10^6

(D) 9×10^{36}

8. *Which ordered pair could NOT lie on the graph of a proportional relationship?*

(A) $(0, 0)$

(B) $(1, 5)$

(C) $(3, 15)$

(D) $(2, 13)$

9. *A line passes through* $(0, -2)$ *and* $(5, 8)$*. What is its equation?*

(A) $y = 2x + 2$

(B) $y = -2x + 2$

(C) $y = 2x - 2$

(D) $y = -2x - 2$

10. *Solve:* $2x + 5y = 20$ *and* $2x + y = 8$.

Your Answer:

11. *If* $f(x) = 2x + 9$, *what is* $f(5)$*?*

Your Answer:

12. Function P is given by the table:

x	0	1	2	3
y	4	7	10	13

Function Q: $y = 2x + 5$. Which function has a greater rate of change?

(A) Function P
(B) Function Q
(C) They have the same rate of change.
(D) Cannot be determined.

13. A table shows $(1,1)$, $(2,8)$, $(3,27)$, $(4,64)$. Is this linear or nonlinear? What pattern do you see?

Your Answer:

14. A gym charges a one-time fee of \$60 plus \$25 per month. What is the total cost after 8 months?

(A) \$200
(B) \$260
(C) \$285
(D) \$480

15. A ball is thrown straight up. Its height over time:

(A) Increases, then decreases — the graph is non-linear.
(B) Increases, then decreases — the graph is linear.
(C) Decreases only — the graph is linear.
(D) Increases only — the graph is nonlinear.

16. Point $N(0,4)$ is rotated $180°$ around the origin. What are the coordinates of N'?

Your Answer:

17. $\triangle ABC \cong \triangle DEF$. If $\angle A = 55°$ and $\angle B = 80°$, what is $\angle F$?

(A) $55°$
(B) $80°$
(C) $45°$
(D) $135°$

18. What is the image of $(-4, 2)$ after a reflection over the y-axis?

(A) $(-4, -2)$
(B) $(4, -2)$
(C) $(4, 2)$
(D) $(2, -4)$

19. Are all squares similar to each other?

(A) No, because they can have different side lengths.
(B) No, because they can have different angles.
(C) Yes, because all squares have equal angles and proportional sides.
(D) Yes, but only if they have the same perimeter.

20. Corresponding angles formed by a transversal and two parallel lines are located:

(A) On opposite sides of the transversal, between the parallel lines
(B) On the same side of the transversal, in matching positions
(C) On opposite sides of the transversal, outside the parallel lines
(D) Adjacent to each other at one intersection

21. A right triangle has legs 10 and 24. What is the hypotenuse?

Your Answer:

22. What is the distance between $(-2, 1)$ and $(4, 9)$?

(A) 10
(B) $\sqrt{10}$
(C) 14
(D) 8

23. A traffic cone has radius 8 in and height 24 in. What is its volume? Use $\pi \approx 3.14$

(A) $4{,}823.0$ in^3
(B) $1{,}607.7$ in^3
(C) 803.8 in^3
(D) 602.9 in^3

24. What is the volume of a square pyramid with base side 6 cm and height 10 cm?

(A) 360 cm^3
(B) 120 cm^3
(C) 180 cm^3
(D) 60 cm^3

25. Which scatter plot pattern suggests a linear association?

(A) Dots that form a U-shape
(B) Dots scattered randomly
(C) Dots that roughly follow a straight line
(D) Dots that form a circle

26. Which describes a "good fit"?

(A) All data points are far from the line.
(B) Data points are close to the line.
(C) All data points are below the line.
(D) The line is vertical.

27. A trend line gives $y = -3x + 45$. Data was collected for $x = 1$ to $x = 10$. Predict y when $x = 5$ and when $x = 30$. State which prediction is more reliable and why.

Your Answer:

28. *A relative frequency table shows the value* 0.15 *in one cell. If the total is* 200, *what is the frequency?*

(A) 15

(B) 30

(C) 10

(D) 20

29. *Draw a tree diagram for flipping a coin twice. List all outcomes and find* $P($*at least one head*$)$.

Your Answer:

30. *You have* 6 *toppings and want to choose* 3 *for a pizza (order doesn't matter). How many ways?*

(A) 120

(B) 20

(C) 18

(D) 216

End of Practice Test 5

Great job finishing the test!

My Score

I got __________ out of 30 questions right.

*Check your answers in the **Answer Key** at the back of the book.*

Review any questions you missed. That's how we learn!

Check Your Score Online!

*Visit **ViewMath Academy** to enter your answers and see which topics you need to review. You can also explore lessons, take quizzes, track your scores, and save your progress!*

viewmath.com/score/8.1.MN.20

*Or go to **viewmath.com/score** and enter code: **8.1.MN.20***

PRACTICE TEST

6

Practice Test 6

30 Questions

Before You Start

- **Read each question carefully** before choosing your answer.
- **Show your work** on scratch paper when you need to.
- **Skip hard questions** and come back to them later.
- **Check your answers** when you're done.
- **Take your time** — there's no rush!

You've Got This!

Do your best and show what you know!

1. Is $\frac{5}{6}$ rational or irrational? Explain.

Your Answer:

2. When converting $0.\overline{123}$ to a fraction, by what power of 10 should you multiply both sides?

(A) 10
(B) 100
(C) 1000
(D) 10000

3. Between which two consecutive integers does $\sqrt{110}$ lie?

(A) 9 and 10
(B) 10 and 11
(C) 11 and 12
(D) 54 and 56

4. A student says $\sqrt{4}+\sqrt{9}=\sqrt{13}$. Is this correct?

(A) Yes, because you add the numbers under the square roots.
(B) No, $\sqrt{4}+\sqrt{9}=2+3=5$, but $\sqrt{13}\approx 3.6$.
(C) Yes, because $4+9=13$.
(D) No, $\sqrt{4}+\sqrt{9}=\sqrt{36}=6$.

5. A cube has a volume of 343 cubic inches. What is the length of one edge?

(A) 7 in.
(B) 49 in.
(C) $114.\overline{3}$ in.
(D) 17 in.

6. What is 9.03×10^{-5} written in standard form?

(A) $903{,}000$
(B) 0.0000903
(C) 0.000903
(D) 0.00903

7. A light year is approximately 9.5×10^{12} km. If a star is 4 light years away, how far is it in kilometers?

(A) 3.8×10^{13} km
(B) 3.8×10^{12} km
(C) 13.5×10^{12} km
(D) 9.5×10^{48} km

8. A graph shows a straight line through $(0, 0)$ and $(2, 9)$. What is the unit rate?

(A) 2
(B) 4.5
(C) 9
(D) 18

9. Which equation represents a line with slope -2 and y-intercept 7?

(A) $y = 7x - 2$
(B) $y = -2x + 7$
(C) $y = 2x + 7$
(D) $y = -2x - 7$

10. On a graph, where is the solution to a system of two linear equations?

(A) At the y-intercept of the first line
(B) At the x-intercept of the second line
(C) At the point where the two lines intersect
(D) At the origin

11. If $f(x) = 7 - x$, what is $f(10)$?

(A) -3
(B) 3
(C) 17
(D) -17

12. The graph below shows Function A and Function B. Which function has a greater rate of change?

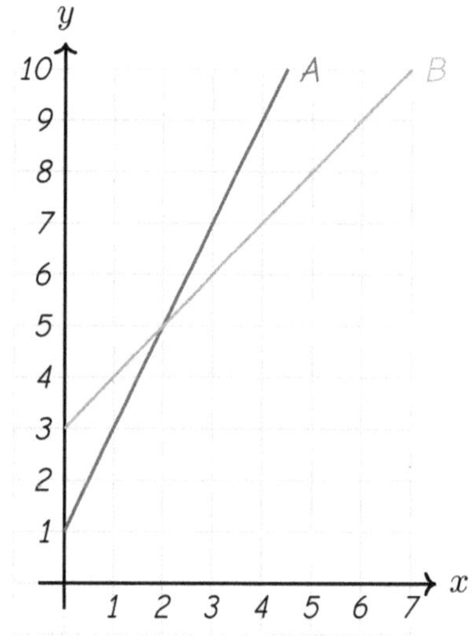

(A) Function A

(B) Function B

(C) They have the same rate of change.

(D) Cannot be determined from the graph.

13. A function increases by 6 for every increase of 1 in x. Is it linear or nonlinear?

Your Answer:

14. A candle is 12 inches tall and burns at 2 inches per hour. What is the function for the candle's height y after x hours?

(A) $y = 2x + 12$

(B) $y = -2x + 12$

(C) $y = 12x - 2$

(D) $y = -12x + 2$

15. A graph shows temperature over a day. From 6 AM to noon, the graph goes up steeply. From noon to 3 PM, it goes up slowly. What can you conclude?

(A) The temperature decreased between noon and 3 PM.

(B) The temperature increased faster in the morning than in the afternoon.

(C) The temperature was constant from noon to 3 PM.

(D) The temperature increased at the same rate all day.

16. A student says that a reflection changes the area of a figure. Is this correct?

(A) Yes, because the figure flips.

(B) Yes, because one dimension reverses.

(C) No, because reflections preserve all measurements.

(D) No, but reflections change the perimeter.

17. Two congruent rectangles each have a perimeter of 28 cm. One has a length of 9 cm. What is its width?

Your Answer:

18. A point (a, b) is translated by $(5, -1)$ to get image $(2, 7)$. What is (a, b)?

Your Answer:

19. Which statement about similar figures is always true?

(A) They have equal side lengths.

(B) They have equal angle measures.

(C) They have the same perimeter.

(D) They have the same area.

20. A triangle has angles $90°$, $x°$, and $(x + 10)°$. Find x.

Your Answer:

21. *A right triangle has legs 1 and 1. What is the exact length of the hypotenuse?*

(A) 2
(B) $\sqrt{2}$
(C) 1
(D) $\sqrt{3}$

22. *Find the distance between $(0, 0)$ and $(2, 3)$. Leave your answer in simplest radical form.*

Your Answer

23. *A cone has radius 3 m and height 12 m. What is its volume in terms of π?*

(A) $108\pi\ m^3$
(B) $36\pi\ m^3$
(C) $12\pi\ m^3$
(D) $324\pi\ m^3$

24. *A triangular pyramid has a base triangle with area $30\ cm^2$ and a height of 9 cm. What is the volume?*

(A) $270\ cm^3$
(B) $90\ cm^3$
(C) $135\ cm^3$
(D) $45\ cm^3$

25. *Which variable typically goes on the x-axis in a scatter plot?*

(A) *The dependent variable*
(B) *The response variable*
(C) *The independent (explanatory) variable*
(D) *The variable with the largest values*

26. *A line passes through $(0, 3)$ and $(4, 11)$. What is the y-intercept?*

(A) 0
(B) 4
(C) 3
(D) 11

27. Which statement about slope in a linear model is true?

(A) A positive slope means y decreases as x increases.

(B) A negative slope means y increases as x increases.

(C) A slope of zero means y stays constant.

(D) The slope is always positive in real data.

28. What is a relative frequency?

(A) The total number of data values

(B) A frequency expressed as a fraction or percentage of a total

(C) The difference between two frequencies

(D) The mode of the data

29. A bag has 3 red, 5 blue, and 2 green marbles. What is $P(\text{blue})$?

(A) $\frac{1}{5}$

(B) $\frac{5}{10}$

(C) $\frac{3}{10}$

(D) $\frac{2}{5}$

30. The Fundamental Counting Principle states that if event A has m outcomes and event B has n outcomes, the total number of outcomes for A followed by B is:

(A) $m + n$

(B) $m \times n$

(C) $m - n$

(D) m^n

★ End of Practice Test 6 ★

Great job finishing the test!

My Score

I got __________ out of 30 questions right.

Check your answers in the ***Answer Key*** *at the back of the book.*

Review any questions you missed. That's how we learn!

Check Your Score Online!

Visit ***ViewMath Academy*** *to enter your answers and see which topics you need to review. You can also explore lessons, take quizzes, track your scores, and save your progress!*

viewmath.com/score/8.1.MN.21

Or go to ***viewmath.com/score*** *and enter code:* ***8.1.MN.21***

PRACTICE TEST

7

Practice Test 7

30 Questions

Before You Start

- **Read each question carefully** before choosing your answer.
- **Show your work** on scratch paper when you need to.
- **Skip hard questions** and come back to them later.
- **Check your answers** when you're done.
- **Take your time** — there's no rush!

You've Got This!

Do your best and show what you know!

1. Give an example of an irrational number between 1 and 2.

Your Answer:

2. A student converts $0.\overline{36}$ and gets $\frac{36}{100}$. What mistake did the student make?

(A) The student divided by 100 instead of 99.

(B) The student forgot to simplify.

(C) The student treated a repeating decimal as a terminating decimal.

(D) The student multiplied by 10 instead of 100.

3. Between which two consecutive integers does $\sqrt{83}$ lie?

Your Answer:

4. A student claims that $2\sqrt{5} = \sqrt{10}$. Is this correct?

(A) Yes, because $2 \times 5 = 10$.

(B) No, $2\sqrt{5} = \sqrt{20}$, not $\sqrt{10}$.

(C) Yes, multiplication distributes into the square root.

(D) No, $2\sqrt{5} = 4\sqrt{5}$.

5. A cube-shaped box has a volume of 216 cubic centimeters. What is the edge length of the box?

Your Answer:

6. Which of the following is 350,000 written in scientific notation?

(A) 35×10^4

(B) 3.5×10^5

(C) 3.5×10^4

(D) 0.35×10^6

7. Earth's mass is about 6×10^{24} kg and Jupiter's mass is about 1.9×10^{27} kg. About how many times more massive is Jupiter than Earth? Round to the nearest whole number.

Your Answer:

8. Which table does NOT represent a proportional relationship?

(A)

x	1	2
y	4	8

(B)

x	2	4
y	5	10

(C)

x	3	6
y	9	15

(D)

x	5	10
y	15	30

9. What is the slope and y-intercept of $y = 5x - 3$?

(A) slope $= -3$, y-intercept $= 5$

(B) slope $= 5$, y-intercept $= -3$

(C) slope $= 3$, y-intercept $= 5$

(D) slope $= 5$, y-intercept $= 3$

10. Solve: $y = -2x + 10$ and $y = x + 1$.

(A) $(2, 3)$

(B) $(3, 4)$

(C) $(4, 5)$

(D) $(5, 0)$

11. If $f(x) = \frac{x+4}{2}$, what is $f(6)$?

Your Answer:

12. *Function R from the previous table has a rate of change of:*

x	0	2	4	6
y	1	9	17	25

(A) 2

(B) 3

(C) 4

(D) 8

13. *The equation* $A = s^2$ *gives the area of a square with side length* s*. Is this function linear or nonlinear?*

(A) *Linear, because area is always positive.*

(B) *Linear, because it has only one variable.*

(C) *Nonlinear, because the variable is squared.*

(D) *Nonlinear, because area is measured in square units.*

14. *A table shows* $(0, 8)$, $(3, 20)$, $(6, 32)$. *What is the equation?*

(A) $y = 3x + 8$

(B) $y = 4x + 8$

(C) $y = 8x + 4$

(D) $y = 6x + 8$

15. *A graph of amount of gas in a car over time goes downward. Is the function increasing, decreasing, or constant?*

Your Answer:

16. *Which transformation flips a figure over a line?*

(A) *Translation*

(B) *Rotation*

(C) *Dilation*

(D) *Reflection*

17. Two figures are congruent if one can be mapped onto the other using which of the following?

(A) Only translations
(B) Only dilations
(C) A sequence of rigid transformations
(D) Any single transformation

18. A point is dilated by factor 3 from the origin. If the image is $(9, -15)$, what was the original point?

(A) $(27, -45)$
(B) $(3, -5)$
(C) $(6, -12)$
(D) $(12, -18)$

19. All circles are similar to each other. Why?

(A) They all have the same radius.
(B) One can always be dilated to match the other.
(C) They all have the same area.
(D) Circles cannot be dilated.

20. In a triangle, the angles are $(2x + 5)°$, $(3x)°$, and $(x + 25)°$. Find x.

Your Answer:

21. A right triangle has hypotenuse 17 and one leg 8. What is the other leg?

(A) 9
(B) 15
(C) 25
(D) $\sqrt{225}$

22. What is the distance between $(1, 1)$ and $(4, 5)$?

(A) 25
(B) 7
(C) $\sqrt{5}$
(D) 5

23. *Find the volume of a cylinder with radius 6 cm and height 5 cm. Leave your answer in terms of π.*

Your Answer:

24. *A rectangular pyramid is shown. What is the volume?*

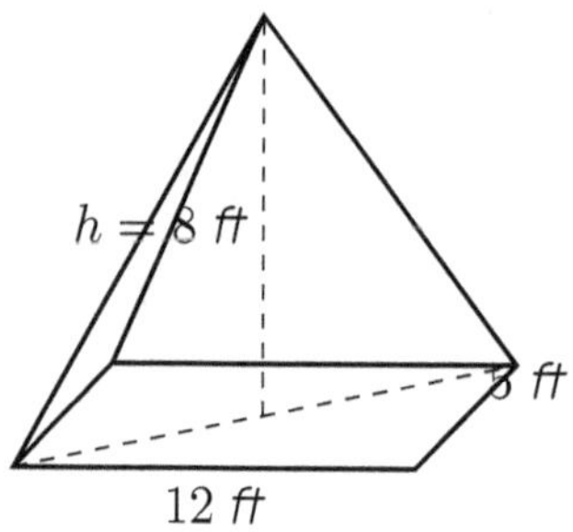

(A) $480\ ft^3$

(B) $160\ ft^3$

(C) $240\ ft^3$

(D) $320\ ft^3$

25. *Does a strong association in a scatter plot prove that one variable causes the other? Explain.*

Your Answer:

26. *A scatter plot shows a strong nonlinear (curved) pattern. Should you draw a straight line of best fit?*

(A) *Yes — always draw a straight line.*

(B) *No — a straight line does not fit curved data well.*

(C) *Yes — but only if there are outliers.*

(D) *No — you should never draw any line.*

27. A model for plant height is $y = 0.4x + 2$ where x is days and y is height in cm. Using the graph, find the predicted height at day 10, and explain what the slope and y-intercept mean.

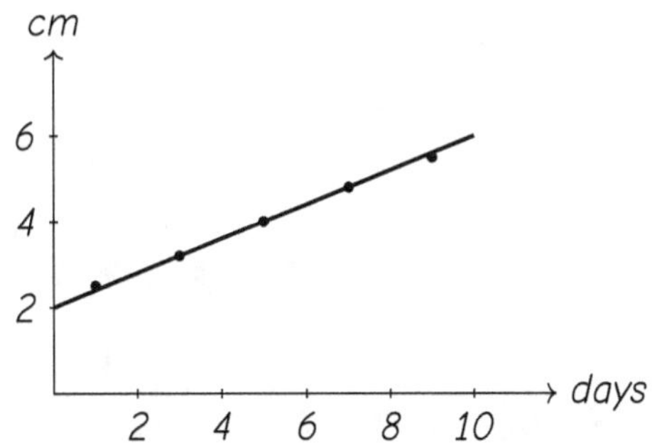

Your Answer:

28. What is a two-way frequency table?

(A) A table with two rows

(B) A table that shows frequencies for two categorical variables

(C) A table that shows only percentages

(D) A table with exactly two columns

29. $P(A) = 0.6$, $P(B) = 0.4$. If A and B are independent, $P(A \text{ and } B) =$?

(A) 1.0

(B) 0.24

(C) 0.20

(D) 0.10

30. A coin is flipped 5 times. How many possible outcomes are there?

(A) 10

(B) 25

(C) 32

(D) 120

★ End of Practice Test 7 ★

Great job finishing the test!

My Score

I got __________ out of 30 questions right.

*Check your answers in the **Answer Key** at the back of the book.*

Review any questions you missed. That's how we learn!

Check Your Score Online!

*Visit **ViewMath Academy** to enter your answers and see which topics you need to review. You can also explore lessons, take quizzes, track your scores, and save your progress!*

viewmath.com/score/8.1.MN.22

*Or go to **viewmath.com/score** and enter code: **8.1.MN.22***

PRACTICE TEST

8

Practice Test 8

30 Questions

Before You Start

- ✓ **Read each question carefully** before choosing your answer.
- ✓ **Show your work** on scratch paper when you need to.
- ✓ **Skip hard questions** and come back to them later.
- ✓ **Check your answers** when you're done.
- ✓ **Take your time** — there's no rush!

★ You've Got This! ★

Do your best and show what you know!

1. Which of the following is a true statement?

(A) Every square root is irrational.

(B) Every integer is irrational.

(C) Every integer is rational.

(D) Every decimal is irrational.

2. The table below shows repeating decimals and their fraction equivalents. Find the missing fraction for $0.\overline{81}$.

Decimal	Unsimplified Fraction	Simplest Form
$0.\overline{27}$	$\frac{27}{99}$	$\frac{3}{11}$
$0.\overline{54}$	$\frac{54}{99}$	$\frac{6}{11}$
$0.\overline{81}$	?	?

Your Answer:

3. Approximate $\sqrt{45}$ to one decimal place.

Your Answer:

4. Which expression equals exactly 6?

(A) $\sqrt{3} \times \sqrt{12}$

(B) $\sqrt{3} + \sqrt{12}$

(C) $\sqrt{6} \times \sqrt{6}$

(D) Both A and C

5. Evaluate $\sqrt[3]{-125}$.

Your Answer:

6. *The bar chart compares the masses of four animals. Which animal's mass is closest to* 1×10^2 *kilograms?*

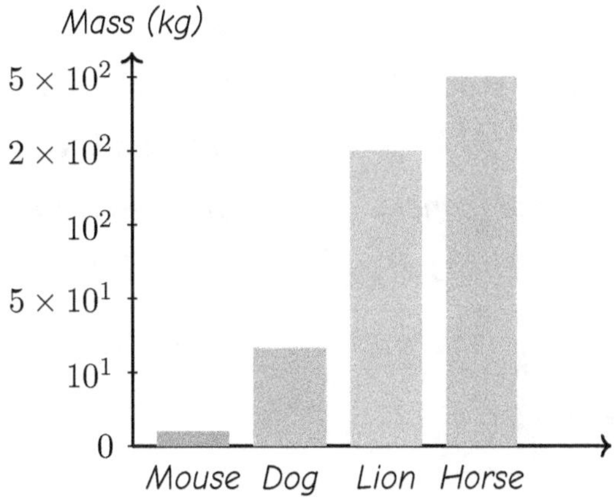

(A) Mouse
(B) Dog
(C) Lion
(D) Horse

7. *A phone stores* 6.4×10^{10} *bytes. A photo uses* 4×10^6 *bytes. How many photos can the phone store?*

(A) 1.6×10^3
(B) 1.6×10^4
(C) 1.6×10^{16}
(D) 2.4×10^4

8. *Machine A fills* $y = 12x$ *bottles per hour. Machine B fills* 50 *bottles in* 5 *hours. Which machine is faster?*

(A) Machine A
(B) Machine B
(C) They fill at the same rate.
(D) Cannot be determined.

9. Using similar triangles, Mia draws a line and picks two different pairs of points. She gets slopes of $\frac{6}{3}$ and $\frac{10}{5}$. What can she conclude?

(A) She made a mistake — the slopes should be different.

(B) The slopes are 2 for both, confirming slope is constant on a line.

(C) The slopes are different because the triangles are different sizes.

(D) Similar triangles cannot be used to find slope.

10. Solve: $x + 2y = 14$ and $x + y = 9$.

Your Answer:

11. The table below shows values for $f(x)$. If $f(x) = 15$, what is x?

x	0	1	2	3	4
$f(x)$	3	6	9	12	15

(A) 2

(B) 3

(C) 4

(D) 5

12. Function R is shown in the table:

x	0	2	4	6
y	1	9	17	25

Function S: $y = 3x + 1$. Which has a greater initial value?

(A) Function R

(B) Function S

(C) They have the same initial value.

(D) Cannot be determined.

13. *Give an example of a nonlinear equation.*

Your Answer:

14. *The table shows a real-world function. Write the equation and explain what m and b represent.*

Hours worked (x)	Pay (y)
0	\$50
1	\$62
2	\$74
3	\$86

Your Answer:

15. *A student fills a glass with water, drinks half, then refills it. Describe the graph of water level vs. time.*

Your Answer:

16. *A point (a, b) is reflected over the x-axis, then reflected over the y-axis. What are the final coordinates?*

Your Answer:

17. $\triangle PQR \cong \triangle STU$. *If* $\angle P = 72°$ *and* $\angle Q = 53°$, *what is* $\angle U$*?*

Your Answer:

18. *Which transformation maps (x, y) to $(-y, x)$?*

(A) Reflection over the x-axis

(B) Reflection over the y-axis

(C) 90° counterclockwise rotation

(D) 180° rotation

19. *Triangle A has sides 6, 8, 10. Triangle B has sides 9, 12, 15. What is the scale factor from A to B?*

Your Answer:

20. *In the figure below, what is the value of y?*

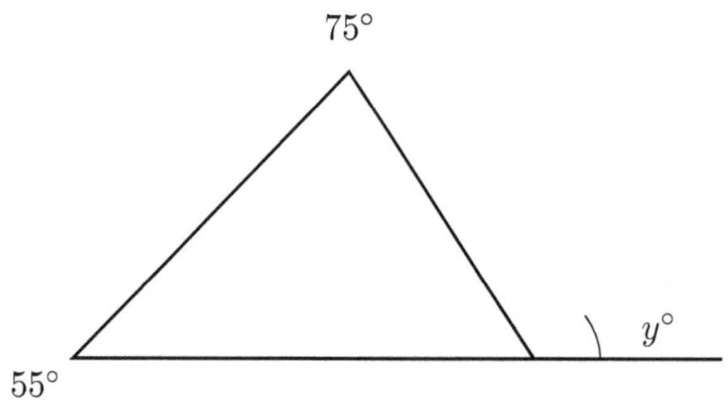

(A) 50°

(B) 105°

(C) 130°

(D) 125°

21. *A rectangular room is 12 ft long and 5 ft wide. What is the diagonal distance across the floor?*

Your Answer:

22. *The distance formula is derived from:*

(A) The area formula for a rectangle

(B) The Pythagorean Theorem

(C) The perimeter formula

(D) The midpoint formula

23. What is the volume of a sphere with radius 3 cm? Use $\pi \approx 3.14$

(A) $28.26\ cm^3$

(B) $113.04\ cm^3$

(C) $36\pi\ cm^3$

(D) $84.78\ cm^3$

24. A square pyramid has volume $48\ in^3$ and height 4 in. What is the base side length?

Your Answer:

25. Can a scatter plot show a positive association that is nonlinear? Give an example.

Your Answer:

26. True or false: The line of best fit must pass through at least one data point.

(A) True — it must go through the middle point.

(B) True — it must go through the first point.

(C) False — it does not need to go through any data point.

(D) False — it must stay above all data points.

27. A model $y = -2x + 50$ represents the number of pages left to read (y) after x hours. How long until the book is finished (all pages read)?

Your Answer:

28.

	Bike	Bus	Total
6th grade	18	22	40
7th grade	25	15	40
Total	43	37	80

What percentage of 6th graders ride the bus?

(A) 45%

(B) 55%

(C) 27.5%

(D) 62.5%

29. A spinner has sections: 30% red, 50% blue, 20% green. It is spun twice. Find P(blue, then green).

Your Answer:

30. How many ways can 3 runners finish a race (1st, 2nd, 3rd) from a group of 8?

(A) 56

(B) 336

(C) 512

(D) 24

★ End of Practice Test 8 ★

Great job finishing the test!

My Score

I got __________ out of 30 questions right.

*Check your answers in the **Answer Key** at the back of the book.*

Review any questions you missed. That's how we learn!

Check Your Score Online!

*Visit **ViewMath Academy** to enter your answers and see which topics you need to review. You can also explore lessons, take quizzes, track your scores, and save your progress!*

viewmath.com/score/8.1.MN.23

*Or go to **viewmath.com/score** and enter code: **8.1.MN.23***

PRACTICE TEST

9

Practice Test 9

30 Questions

Before You Start

- **Read each question carefully** before choosing your answer.
- **Show your work** on scratch paper when you need to.
- **Skip hard questions** and come back to them later.
- **Check your answers** when you're done.
- **Take your time** — there's no rush!

You've Got This!

Do your best and show what you know!

1. *Which of the following could NOT be the decimal expansion of a rational number?*

(A) 2.500

(B) $0.\overline{9}$

(C) $1.41421356\ldots$ *(non-repeating)*

(D) $0.\overline{36}$

2. *What is $0.\overline{4}$ written as a fraction?*

(A) $\frac{4}{10}$

(B) $\frac{4}{9}$

(C) $\frac{2}{5}$

(D) $\frac{1}{4}$

3. *A student says $\sqrt{45}$ is between 6 and 7. Is the student correct?*

(A) *Yes, because $6^2 = 36$ and $7^2 = 49$.*

(B) *No, it is between 5 and 6.*

(C) *No, it is between 7 and 8.*

(D) *No, $\sqrt{45} = 9$ exactly.*

4. *Which is larger: $\sqrt{2} + \sqrt{3}$ or $\sqrt{10}$?*

(A) $\sqrt{2} + \sqrt{3}$*, because $2 + 3 = 5 > \sqrt{10}$*

(B) $\sqrt{10}$*, because $10 > 5$*

(C) $\sqrt{10}$*, because $\sqrt{10} \approx 3.16 > 3.15 \approx \sqrt{2} + \sqrt{3}$*

(D) *They are equal.*

5. *Between which two consecutive whole numbers does $\sqrt{50}$ lie?*

(A) *6 and 7*

(B) *7 and 8*

(C) *24 and 26*

(D) *8 and 9*

6. *Which is larger: 8×10^5 or 3×10^6?*

(A) 8×10^5*, because $8 > 3$*

(B) 3×10^6*, because the exponent is larger*

(C) *They are equal*

(D) *It cannot be determined*

7. *Simplify* $(7 \times 10^{-3})(8 \times 10^{-2})$. *Write in proper scientific notation.*

(A) 56×10^{-5}

(B) 5.6×10^{-5}

(C) 5.6×10^{-4}

(D) 5.6×10^{6}

8. *Two delivery services are compared. Service P delivers* $y = 4x$ *packages per hour. Service Q's data is shown in the table.*

Hours (x)	2	4	6	8
Packages (y)	10	20	30	40

How many more packages does Service Q deliver than Service P in 10 *hours?*

Your Answer:

9. *Which line passes through the origin?*

(A) $y = 2x + 1$

(B) $y = -3x$

(C) $y = x - 5$

(D) $y = 4x + 4$

10. *Use the tables to find the solution to the system of equations.*

Equation 1: $y = 2x + 1$

x	y
0	1
1	3
2	5
3	7
4	9

Equation 2: $y = -x + 7$

x	y
0	7
1	6
2	5
3	4
4	3

Your Answer:

11. If $f(x) = -x + 8$, what is $f(-3)$?

(A) 5
(B) -5
(C) 11
(D) -11

12. Function A: $y = 5x + 3$. Function B: $y = 2x + 10$. Which function has a greater rate of change?

(A) Function A
(B) Function B
(C) They have the same rate of change.
(D) Cannot be determined.

13. Which of these is NOT a linear function?

(A) $y = 0$
(B) $y = 100x$
(C) $y = -x + 5$
(D) $y = \sqrt{x}$

14. A linear function passes through $(0, 5)$ and $(2, 11)$. What is its equation?

(A) $y = 2x + 5$
(B) $y = 3x + 5$
(C) $y = 5x + 3$
(D) $y = 6x + 5$

15. A decreasing graph does NOT necessarily mean:

(A) The output values are getting smaller.
(B) The graph goes downward from left to right.
(C) The output values are negative.
(D) The rate of change is negative.

16. After a translation, a triangle's longest side was originally 8 cm. What is the longest side of the image?

(A) 4 cm
(B) 8 cm
(C) 16 cm
(D) It depends on the direction of the translation.

17. A pentagon is translated 4 units left and then rotated 90°. Is the image congruent to the original?

(A) Yes
(B) No, because two transformations were used
(C) No, because it was rotated
(D) Only if it is a regular pentagon

18. Point $(-3, 5)$ is rotated 90° counterclockwise around the origin. What is its image?

(A) $(5, 3)$
(B) $(-5, -3)$
(C) $(3, -5)$
(D) $(-5, 3)$

19. A dilation with $k = 1$ produces:

(A) A figure twice as large
(B) A congruent figure
(C) A figure half as large
(D) No figure at all

20. Triangle ABC has $\angle A = 45^\circ$ and $\angle B = 85^\circ$. What type of triangle is it?

(A) Acute
(B) Right
(C) Obtuse
(D) Equilateral

21. Is the triangle with sides 9, 40, 41 a right triangle?

(A) No, because $9 + 40 \neq 41$.
(B) No, because $9^2 + 40^2 \neq 41^2$.
(C) Yes, because $9^2 + 40^2 = 41^2$.
(D) Yes, because $41 - 40 = 1$.

22. What is the distance between $(0, 0)$ and $(3, 4)$?

(A) 7
(B) 5
(C) $\sqrt{7}$
(D) 12

23. *A cone has volume 80π cm^3 and radius 4 cm. What is the height?*

(A) 5 *cm*

(B) 15 *cm*

(C) 10 *cm*

(D) 20 *cm*

24. *A rectangular pyramid has a base of 8×5 m and a height of 12 m. What is its volume?*

(A) 480 m^3

(B) 160 m^3

(C) 240 m^3

(D) 320 m^3

25. *What type of association does this scatter plot show?*

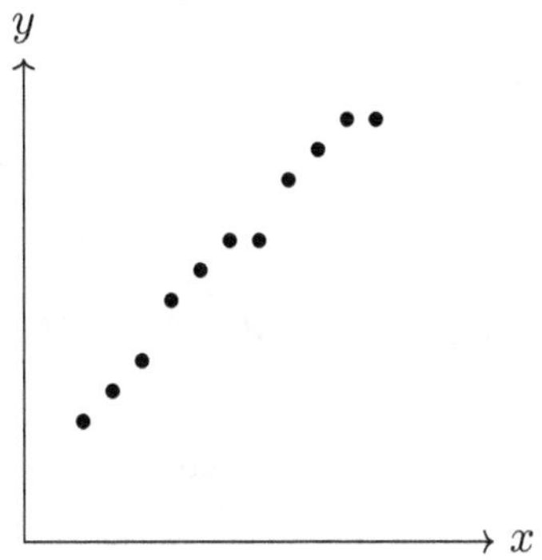

(A) *Negative linear*

(D) *No association*

(C) *Positive linear*

(D) *Nonlinear*

26. *A trend line has equation $y = 1.5x + 2$. Using this, what is y when $x = 4$?*

(A) 6

(B) 7

(C) 8

(D) 10

27. *A scatter plot shows data and a trend line. The actual point at $x = 3$ is at $y = 8$, but the trend line passes through $(3, 6)$. What is the residual?*

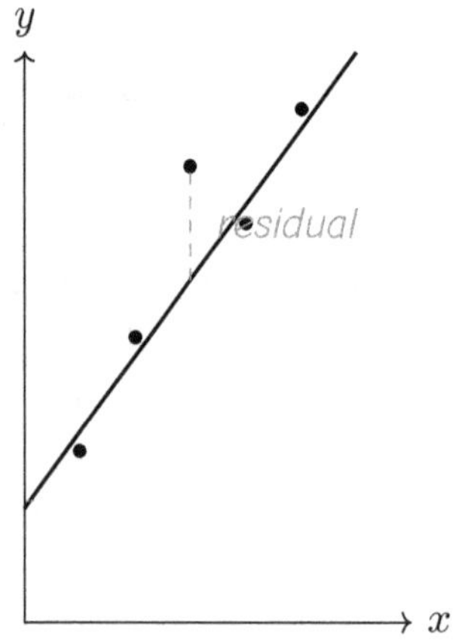

(A) -2

(B) 2

(C) 6

(D) 14

28. *Using the same table, what percentage of students who did NOT study passed?*

(A) 40%

(B) 60%

(C) 25%

(D) 50%

29. *A coin is flipped 3 times. What is P(all heads)?*

(A) $\frac{3}{8}$

(B) $\frac{1}{8}$

(C) $\frac{1}{3}$

(D) $\frac{1}{6}$

30. *A password must be 4 characters: first 2 are letters (A–Z, no repeat), last 2 are digits (0–9, no repeat). How many passwords?*

Your Answer:

★ End of Practice Test 9 ★

Great job finishing the test!

My Score

I got __________ out of 30 questions right.

*Check your answers in the **Answer Key** at the back of the book.*

Review any questions you missed. That's how we learn!

Check Your Score Online!

*Visit **ViewMath Academy** to enter your answers and see which topics you need to review. You can also explore lessons, take quizzes, track your scores, and save your progress!*

viewmath.com/score/8.1.MN.24

*Or go to **viewmath.com/score** and enter code: **8.1.MN.24***

PRACTICE TEST

10

Practice Test 10

30 Questions

Before You Start

- **Read each question carefully** before choosing your answer.
- **Show your work** on scratch paper when you need to.
- **Skip hard questions** and come back to them later.
- **Check your answers** when you're done.
- **Take your time** — there's no rush!

You've Got This!

Do your best and show what you know!

1. Look at the Venn diagram below. In which region does $\sqrt{5}$ belong?

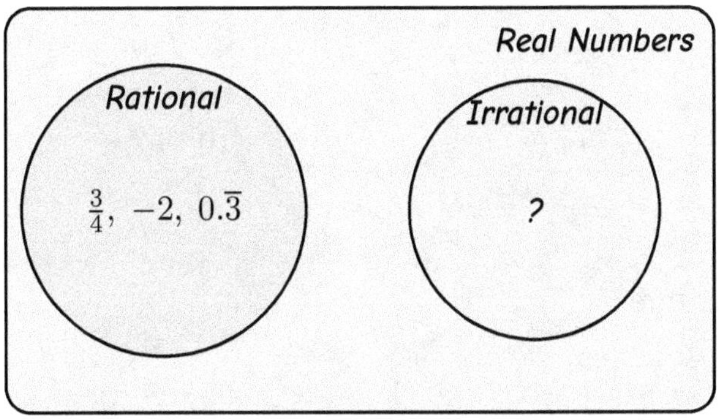

(A) Rational region, because $\sqrt{5} = 2.5$

(B) Irrational region, because 5 is not a perfect square

(C) Rational region, because $\sqrt{5}$ is a square root

(D) Neither region, because $\sqrt{5}$ is not a real number

2. The steps below show a student's work for converting $0.\overline{2}$ to a fraction. Which step contains the error?

Step	Work
1	Let $x = 0.222\ldots$
2	$10x = 2.222\ldots$
3	$10x - x = 2.222\ldots - 0.222\ldots$
4	$9x = 2$
5	$x = \frac{2}{10}$

(A) Step 2

(B) Step 3

(C) Step 4

(D) Step 5

3. Between which two consecutive integers does $\sqrt{50}$ lie?

(A) 6 and 7

(B) 7 and 8

(C) 8 and 9

(D) 24 and 26

4. *Order from least to greatest:* π, $\sqrt{10}$, 3.2.

Ⓐ $\pi,\ 3.2,\ \sqrt{10}$

Ⓑ $3.2,\ \pi,\ \sqrt{10}$

Ⓒ $\sqrt{10},\ \pi,\ 3.2$

Ⓓ $\pi,\ \sqrt{10},\ 3.2$

5. *Which statement about* $\sqrt{2}$ *is true?*

Ⓐ $\sqrt{2}$ *is a rational number*

Ⓑ $\sqrt{2}$ *is an integer*

Ⓒ $\sqrt{2}$ *is an irrational number*

Ⓓ $\sqrt{2} = 1.5$

6. *The diagram below shows three cells viewed under a microscope, with their actual sizes labeled. Write all three sizes in scientific notation, then list them from smallest to largest.*

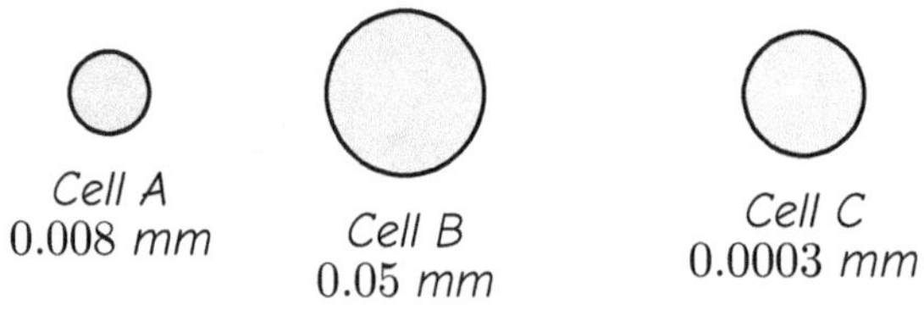

Your Answer:

7. *Compute* $\frac{3.6\times10^{10}}{1.2\times10^{4}}$. *Write your answer in scientific notation.*

Your Answer:

8. *If* $y = kx$ *and* $y = 54$ *when* $x = 6$*, find* y *when* $x = 11$.

Your Answer:

9. *A line passes through the origin and* $(3, 12)$. *What is its equation?*

(A) $y = 3x$

(B) $y = 12x$

(C) $y = 4x$

(D) $y = \frac{1}{4}x$

10. *Solve using elimination:* $x + y = 10$ *and* $x - y = 4$.

(A) $(5, 5)$

(B) $(7, 3)$

(C) $(6, 4)$

(D) $(8, 2)$

11. *Use the graph of* f *below to find* $f(1) + f(4)$.

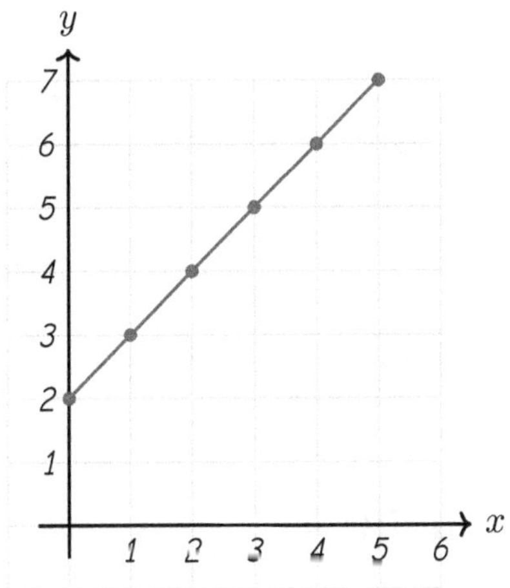

Your Answer:

12. *Function A:* $y = 2x + 100$. *Function B:* $y = 8x + 10$. *At what value of* x *do the functions have the same output?*

(A) $x = 10$

(B) $x = 15$

(C) $x = 18$

(D) $x = 20$

13. *The table below shows a function. Calculate the first differences in y and determine whether the function is linear or nonlinear.*

x	0	1	2	3	4
y	-3	1	5	9	13

Your Answer:

14. *Sam has $\$120$ and spends $\$15$ per day. Which function models the money remaining after x days?*

(A) $y = 15x + 120$
(B) $y = -15x + 120$
(C) $y = 120x - 15$
(D) $y = -120x + 15$

15. *Two distance-time graphs start at the same point. Graph A is steeper than Graph B. What can you say?*

(A) *Object A is farther from the start.*
(B) *Object A is traveling faster than Object B.*
(C) *Object B is traveling faster than Object A.*
(D) *Both objects travel at the same speed.*

16. *Point $M(5, -3)$ is reflected over the y-axis. What are the coordinates of M'?*

Your Answer:

17. *A student says two circles with radius 5 cm are always congruent. Is this correct?*

(A) *No, circles cannot be congruent.*
(B) *No, they need the same center too.*
(C) *Yes, because they have the same radius.*
(D) *Yes, but only if they are in the same position.*

18. Point M is shown on the coordinate plane at $(-2, 5)$. Find the coordinates of M' after a $90°$ counterclockwise rotation around the origin.

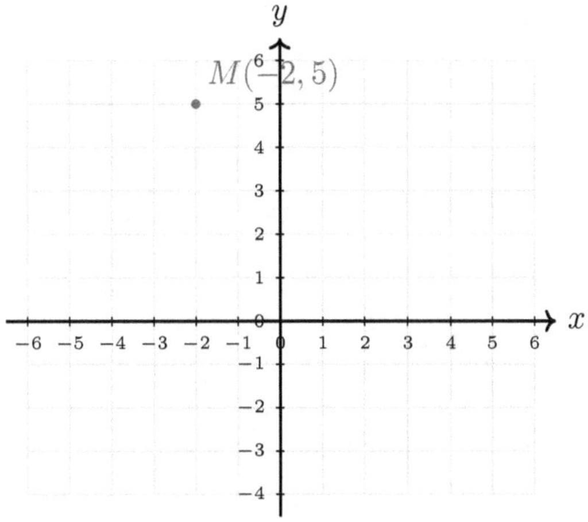

Your Answer:

19. Two similar triangles have a scale factor of $\frac{1}{3}$ from the larger to the smaller. If the larger triangle's perimeter is 36 cm, what is the smaller triangle's perimeter?

(A) 12 cm

(B) 108 cm

(C) 9 cm

(D) 18 cm

20. Two parallel lines are cut by a transversal. If one co-interior (same-side interior) angle is $135°$, what is the other?

(A) $135°$

(B) $45°$

(C) $55°$

(D) $225°$

21. *In a right triangle, the legs are* 6 *and* 8. *What is the hypotenuse?*

Ⓐ 14
Ⓑ 10
Ⓒ 48
Ⓓ $\sqrt{48}$

22. *Two points are at* $(3, 7)$ *and* $(3, -5)$. *What is the distance between them?*

Ⓐ 2
Ⓑ 12
Ⓒ -12
Ⓓ $\sqrt{12}$

23. *A cylindrical can has radius* 3 *in and height* 8 *in. How much soup can it hold? Use* $\pi \approx 3.14$.

Your Answer

24. *A square pyramid and a rectangular prism share the same base* (6×6 *in*) *and height* (10 *in*). *What is the volume of the pyramid?*

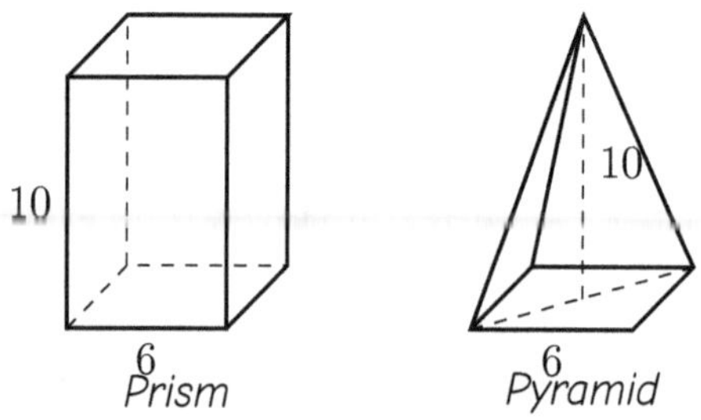

Ⓐ $360\ in^3$
Ⓑ $180\ in^3$
Ⓒ $120\ in^3$
Ⓓ $60\ in^3$

25. True or false: A scatter plot can show both clustering and outliers at the same time.

(A) True — they describe different features of the data.

(B) False — data has either clusters or outliers, not both.

(C) True — but only if there is no association.

(D) False — outliers are always in clusters.

26. Why should outliers generally be ignored when drawing a line of best fit?

Your Answer:

27. A linear model is $y = 3x + 10$, where x is hours studied and y is the test score. What does the y-intercept represent?

(A) The score after 3 hours of studying

(B) The score with 0 hours of studying

(C) The number of hours needed to pass

(D) The maximum possible score

28. The bar chart below shows survey results about favorite season by gender. Which statement is supported?

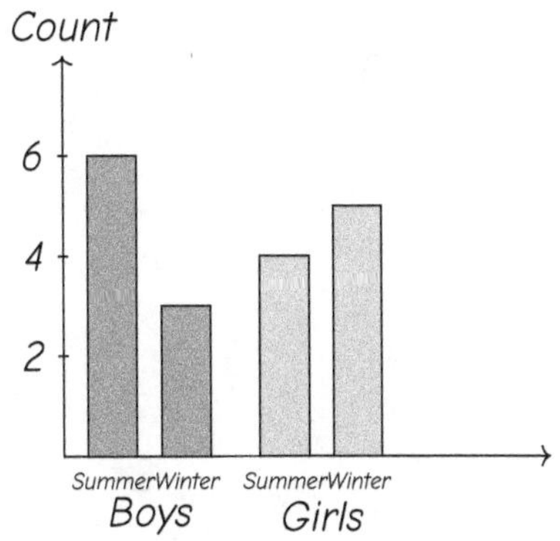

(A) Boys and girls prefer summer equally.

(B) Boys prefer summer more, girls prefer winter more.

(C) Girls prefer summer more than boys.

(D) There is no difference in preference.

29. *A fair die is rolled twice. What is P(both rolls are 6)?*

(A) $\frac{1}{6}$

(B) $\frac{1}{12}$

(C) $\frac{1}{36}$

(D) $\frac{2}{6}$

30. *You pick 2 books from a shelf of 5 to read (order doesn't matter). How many combinations?*

(A) 20

(B) 10

(C) 25

(D) 5

End of Practice Test 10

Great job finishing the test!

My Score

I got __________ out of 30 questions right.

*Check your answers in the **Answer Key** at the back of the book.*

Review any questions you missed. That's how we learn!

Check Your Score Online!

*Visit **ViewMath Academy** to enter your answers and see which topics you need to review. You can also explore lessons, take quizzes, track your scores, and save your progress!*

viewmath.com/score/8.1.MN.25

*Or go to **viewmath.com/score** and enter code: **8.1.MN.25***

Answer Key & Explanations

Answer Key

First try each test on your own, then check your work here.

Practice Test 1 — Answer Key

1 False
2 C
3 D
4 B
5 B
6 $\approx 3.1 \times 10^{12}$
7 C
8 B
9 $y = 10x + 25$
10 $(9, 6)$
11 C
12 Function A: rate $= 3$, initial $= 20$. Function B: rate $= 5$, initial $= 5$. At $x = 10$, $B = 55$ is greater than $A = 50$.
13 Yes
14 B
15 Constant
16 C
17 Reflection over the y-axis
18 $(-5, -2)$
19 B
20 B
21 B
22 C
23 A
24 C
25 B
26 A
27 B
28 Boys: $\frac{20}{30} \approx 66.7\%$. Girls: $\frac{12}{20} = 60\%$. The percentages are close, so association is weak or none.
29 B
30 Choosing 2 items to include is the same as choosing 3 items to exclude. $\binom{n}{r} = \binom{n}{n-r}$.

Time to Learn!

*Review the explanations below, **especially for the questions you missed.***

Understanding why each answer is correct builds stronger problem-solving skills.

***Tip:** Circle any questions you got wrong, then read their explanation carefully.*

Practice Test 1 — Detailed Explanations

1 $\frac{22}{7} \approx 3.142857\ldots$ *is a rational approximation of* π*, but* $\pi = 3.14159265\ldots$ *is irrational. They are close but not equal.*

2 $0.\overline{123}$ *has a 3-digit repeating block, so you should multiply by* $10^3 = 1000$*, not* 100*.*

3 $3.74^2 = 13.9876$ *(under) and* $3.75^2 = 14.0625$ *(just over).* 14.0625 *is very close to* 14*, so* 3.75 *is the best next estimate.*

4 $\sqrt{7} \approx 2.646$*, so* $5 - \sqrt{7} \approx 5 - 2.646 = 2.354 \approx 2.4$*.*

5 *If* $s = 8$*, then* $A = 8^2 = 64$*. If* $e = 4$*, then* $V = 4^3 = 64$*. Since* $A = V = 64$*, the pair* $s = 8, e = 4$ *works.*

6 $498{,}000{,}000 \approx 5 \times 10^8$ *and* $6{,}200 \approx 6.2 \times 10^3$*. Product* $\approx 5 \times 6.2 \times 10^{11} = 31 \times 10^{11} = 3.1 \times 10^{12}$*.*

7 $(3 \times 10^4)(5 \times 10^3) = 15 \times 10^7 = 1.5 \times 10^8$*.*

8 *At* $\$8$ *per hour with no extra fee, the relationship is proportional:* $y = 8x$*.*

9 *Slope (rate)* $= 10$ *and* y*-intercept (initial fee)* $= 25$*:* $y = 10x + 25$*.*

10 *Add:* $2x = 18$*,* $x = 9$*. Then* $y = 15 - 9 = 6$*.*

11 $g(x) = 0$ *where the line crosses the* x*-axis. The graph crosses at* $x = 4$*.*

12 A*: slope 3,* $b = 20 \Rightarrow A(10) = 3(10) + 20 = 50$*.* B*: slope 5,* $b = 5 \Rightarrow B(10) = 5(10) + 5 = 55$*. Function B is greater at* $x = 10$*.*

13 $y = -12$ is a horizontal line, which can be written as $y = 0x + (-12)$ ($m = 0$, $b = -12$). It is linear.

14 The rate is \$3 per mile ($m = 3$) and the flat fee is \$4 ($b = 4$). So $y = 3x + 4$.

15 A straight line has a constant slope, meaning the rate of change is the same everywhere along the line.

16 A translation moves the figure to a new position. Side lengths, angle measures, and parallelism are all preserved.

17 Each x-coordinate is negated while y stays the same: $(1,1) \to (-1,1)$, etc. This is a reflection over the y-axis.

18 $180°$ rotation: $(x,y) \to (-x,-y)$. So $(5,2) \to (-5,-2)$.

19 Divide the image coordinates by the original: $\frac{12}{4} = 3$ and $\frac{-6}{-2} = 3$. The scale factor is 3.

20 Alternate interior angles are equal: $4x = 80$, so $x = 20$.

21 $c^2 = 5^2 + 5^2 = 25 + 25 = 50$, so $c = \sqrt{50}$.

22 $d = \sqrt{(6-2)^2 + (-2-(-5))^2} = \sqrt{16+9} = \sqrt{25} = 5$.

23 A cone is exactly $\frac{1}{3}$ the volume of a cylinder with the same base and height: $\frac{450\pi}{3} = 150\pi$ cm^3.

24 Prism volume $= 96$ cm^3. Pyramid volume $= \frac{1}{3}(16)(6) = 32$ cm^3. $96/32 = 3$.

25 One variable goes up while the other goes down, forming a negative association.

26 Slope $= \frac{7-3}{5-1} = 1$. y-intercept: $3 = 1(1) + b$, $b = 2$. Equation: $y = x + 2$.

27 The linear pattern observed in data may not hold outside the data range. Real-world relationships can change.

28 Compare the conditional relative frequencies. Since 66.7% and 60% are similar, the preference doesn't change much by gender.

29 $P(\text{not event}) = 1 - P(\text{event}) = 1 - \frac{3}{4} = \frac{1}{4}$.

30 $\binom{5}{2} = 10$ and $\binom{5}{3} = 10$. Each way of selecting 2 items corresponds to leaving out 3 items.

Practice Test 2 — Answer Key

1 C

2 Multiply by 100; $100x = 54.\overline{54}$; $99x = 54$; $x = \frac{54}{99} = \frac{6}{11}$

3 A: ≈ 4.2, B: ≈ 6.3, C: ≈ 8.5

4 C

5 B

6 B

7 A

8 180 mL

9 A

10 $(3, 1)$

11 B

12 7

13 Nonlinear

14 4

15 C

16 C

17 B

18 A

19 108 cm^2

20 B

21 B

22 9

23 C

24 100 cm^3

25 C

26 Approximate line through $(1, 2)$ and $(7, 9)$: slope $\approx \frac{7}{6} \approx 1.17$.

27 B

28 D

29 C

30 B

Time to Learn!

Review the explanations below, **especially for the questions you missed.**

Understanding why each answer is correct builds stronger problem-solving skills.

Tip: Circle any questions you got wrong, then read their explanation carefully.

Practice Test 2 — Detailed Explanations

1 $P = \sqrt{2}$ is irrational (since 2 is not a perfect square) and $Q = 3$ is an integer, which is rational.

2 The repeating block has 2 digits, so multiply by 100. Then $100x - x = 99x = 54$, giving $x = \frac{54}{99} = \frac{6}{11}$.

3 Side $= \sqrt{Area}$. $\sqrt{18} \approx 4.24 \approx 4.2$. $\sqrt{40} \approx 6.32 \approx 6.3$. $\sqrt{72} \approx 8.49 \approx 8.5$.

4 $\sqrt{5} \approx 2.236$ and $\sqrt{3} \approx 1.732$. Adding: $2.236 + 1.732 = 3.968 \approx 4.0$.

5 $x = \pm\sqrt{\frac{9}{16}} = \pm\frac{3}{4}$, since $\left(\frac{3}{4}\right)^2 = \frac{9}{16}$.

6 $\frac{4.2\times10^3}{4.2\times10^{-3}} = 10^{3-(-3)} = 10^6$. So 4.2×10^3 is 10^6 (one million) times as large.

7 Divide the coefficients: $\frac{8}{2} = 4$. Subtract the exponents: $10^{7-3} = 10^4$. Answer: 4×10^4.

8 Rate $= \frac{45}{15} = 3$ mL/min. In 60 minutes: $3 \times 60 = 180$ mL.

9 $m = \frac{11-2}{4-1} = 3$. Using $(1, 2)$. $2 = 3(1) + b$, so $b = -1$. Equation: $y = 3x - 1$.

10 Substitute: $3x + (x - 2) = 10$, so $4x = 12$, $x = 3$. Then $y = 3 - 2 = 1$.

11 $f(2) = 10 - 3(2) = 10 - 6 = 4$ pieces remaining.

12 The initial value is the output when $x = 0$, which is 7.

13 Check the y-differences: $2-1=1$, $4-2=2$, $5-4=1$, $7-5=2$. The differences are not all the same $(1, 2, 1, 2)$, so the rate of change is not constant. The function is nonlinear.

14 $\frac{25-9}{6-2} = \frac{16}{4} = 4$.

15 A straight line means linear. Going upward means increasing.

16 Reflecting over the y-axis flips the sign of the x-coordinate while keeping y the same: $(7, -2) \to (-7, -2)$.

17 Equal angles make triangles similar, but not necessarily congruent — their sides could be different lengths.

18 First translate: $(1+3, 4+(-2)) = (4, 2)$. Then reflect over the y-axis: $(4, 2) \to (-4, 2)$.

19 Area scales by k^2: $12 \times 3^2 = 12 \times 9 = 108$ cm^2.

20 Alternate interior angles formed by parallel lines and a transversal are equal: $72°$.

21 $c^2 = 8^2 + 15^2 = 64 + 225 = 289$, so $c = \sqrt{289} = 17$.

22 Same x-coordinates, so distance $= |7-(-2)| = 9$.

23 $V = \pi r^2 h = \pi(49)(3) = 147\pi$ m^3.

24 Base area $= \frac{1}{2}(5)(12) = 30$ cm^2. $V = \frac{1}{3}(30)(10) = 100$ cm^3.

25 Point P at $(3, 7)$ is far above the general upward trend around $y \approx 2.5$ when $x = 3$. It is the outlier.

26 $m = \frac{9-2}{7-1} = \frac{7}{6} \approx 1.17$. The line rises about 1.2 units for each 1-unit increase in x.

27 *In $y = mx + b$, the slope m represents the rate of change: how much y changes for each 1-unit increase in x.*

28 *Joint frequencies fill the interior cells: $3 \times 4 = 12$.*

29 *2 coin outcomes $\times$ 6 die outcomes $= 12$ total outcomes.*

30 *$3! = 6$. The tree shows all 6 arrangements: ABC, ACB, BAC, BCA, CAB, CBA.*

Practice Test 3 — Answer Key

#	Answer	#	Answer	#	Answer
1	C	11	B	21	B
2	C	12	B	22	10
3	B	13	B	23	C
4	C	14	$y = 4x + 4$	24	A
5	C	15	B	25	C
6	A	16	B	26	Approximately 2 to 2.25
7	A	17	A	27	B
8	A	18	A	28	B
9	C	19	C	29	B
10	D	20	B	30	B

Time to Learn!

*Review the explanations below, **especially for the questions you missed**.*

Understanding why each answer is correct builds stronger problem-solving skills.

Tip: *Circle any questions you got wrong, then read their explanation carefully.*

Practice Test 3 — Detailed Explanations

1 $0.\overline{81}$ *is a repeating decimal, so it can be written as a fraction:* $0.\overline{81} = \frac{81}{99} = \frac{9}{11}$. *The others are irrational.*

2 *Let* $x = 0.\overline{27}$. *Then* $100x = 27.\overline{27}$. *Subtract:* $99x = 27$, *so* $x = \frac{27}{99} = \frac{3}{11}$.

3 $7.7^2 = 59.29$ *and* $7.8^2 = 60.84$. *So* $\sqrt{60} \approx 7.75$, *making* 8 *a less accurate estimate.* 7.7 *or* 7.8 *is better.*

4 $4\sqrt{2} \approx 4 \times 1.414 = 5.656$ *and* $\sqrt{30} \approx 5.477$. *Since* $5.656 > 5.477$, $4\sqrt{2}$ *is larger.*

5 *Side* $= \sqrt{196} = 14$ *cm, since* $14 \times 14 = 196$.

6 *Move the decimal* 4 *places right:* $0.00072 = 7.2 \times 10^{-4}$.

7 $(9.5 \times 10^{-18})(2 \times 10^{6}) = 19 \times 10^{-12} = 1.9 \times 10^{-11}$ *g.*

8 *Store A:* $k = 2$ *per pound. Store B:* $k = \frac{7.50}{3} = 2.50$ *per pound. Store A is cheaper.*

9 *Steepness is* $|m|$: $|1| = 1$, $|3| = 3$, $|-5| = 5$, $|\frac{1}{2}| = 0.5$. *The steepest has* $|m| = 5$.

10 *Rewrite the second:* $-2y = -4x - 10$, *so* $y = 2x + 5$. *Both equations are the same line. Infinitely many solutions.*

11 $f(0) = -1$ *and* $f(2) = 7$. *So* $f(0) + f(2) = -1 + 7 = 6$.

12 $|-4| = 4 > |-1| = 1$, *so Function B decreases faster.*

13 $7 - 3 = 4$, $11 - 7 = 4$, $15 - 11 = 4$. *The constant change of* 4 *proves this is linear.*

14 *Slope:* $\frac{24-4}{5-0} = \frac{20}{5} = 4$. y*-intercept:* 4. *Equation:* $y = 4x + 4$.

15 A downward straight line means the quantity decreases at a constant rate — water is leaking steadily.

16 The $90°$ CCW rule is $(x, y) \to (-y, x)$. So $(-1, 4) \to (-4, -1)$.

17 Each vertex of Figure A is shifted 6 units right to get Figure B: $(1,1) \to (7,1)$, $(4,1) \to (10,1)$, etc. Since a translation preserves size and shape, the quadrilaterals are congruent.

18 $90°$ CCW: $(x, y) \to (-y, x)$. So $(0, -5) \to (5, 0)$.

19 Multiply the side by the scale factor: $3 \times 4 = 12$ cm.

20 $180 - 54 - 54 = 72°$.

21 The diagonal of a square with side s is $s\sqrt{2}$. So the distance is $90\sqrt{2} \approx 127.3$ ft.

22 Horizontal: $4 - (-4) = 8$. Vertical: $4 - (-2) = 6$. $d = \sqrt{8^2 + 6^2} = \sqrt{64 + 36} = \sqrt{100} = 10$.

23 $V = \frac{4}{3}\pi(3r)^3 = \frac{4}{3}\pi(27r^3) = 27 \times \frac{4}{3}\pi r^3$. The volume is 27 times as large.

24 $V = \frac{1}{3}(49)(12) = \frac{588}{3} = 196$ cm^3.

25 A strong positive trend corresponds to an r-value close to 1. $r = 0.9$ indicates a strong positive association.

26 Using endpoints: $m \approx \frac{11-2}{5-1} = \frac{9}{4} = 2.25$. A slope around 2 is a reasonable estimate.

27 Residual = actual − predicted. A positive residual means the actual value is above the predicted value.

28 *The 20 is the count for a specific combination of two variables (7th grade AND soccer), making it a joint frequency.*

29 *Independent events have no effect on each other. The occurrence of one does not change the probability of the other.*

30 $4 \times 3 \times 2 = 24$. *Order matters and no repeats.*

Practice Test 4 — Answer Key

#	Answer	#	Answer	#	Answer
1	D	11	C	21	B
2	$\frac{53}{90}$	12	$A = 21,\ B = 21$	22	C
3	C	13	C	23	B
4	6	14	$y = 3x$	24	70 cm^3
5	B	15	C	25	D
6	C	16	B	26	B
7	C	17	C	27	B
8	Taxi B	18	B	28	B
9	B	19	A	29	$P(\text{both blue}) = \frac{6}{10} \times \frac{5}{9} = \frac{30}{90} = \frac{1}{3}$.
10	B	20	A	30	B

Time to Learn!

Review the explanations below, ***especially for the questions you missed.***

Understanding why each answer is correct builds stronger problem-solving skills.

Tip: *Circle any questions you got wrong, then read their explanation carefully.*

Practice Test 4 — Detailed Explanations

1 $\sqrt{7}$ is irrational because 7 is not a perfect square. The other choices are all rational: $\frac{5}{8} = 0.625$, $0.\overline{6} = \frac{2}{3}$, and $\sqrt{9} = 3$.

2 Let $x = 0.5888\ldots$. Then $10x = 5.888\ldots$ and $100x = 58.888\ldots$. Subtract: $90x = 53$, so $x = \frac{53}{90}$.

3 $3^2 = 9$ and $4^2 = 16$. Since $9 < 15 < 16$, we have $3 < \sqrt{15} < 4$. Therefore $\sqrt{15} > 3$ is true. In fact, $\sqrt{15} \approx 3.87$.

4 $\sqrt{2} \times \sqrt{18} = \sqrt{2 \times 18} = \sqrt{36} = 6$.

5 $7 \times 7 = 49$, so $\sqrt{49} = 7$.

6 Move the decimal 3 places right: $6.1 \times 10^3 = 6{,}100$.

7 $\frac{2 \times 10^{-6}}{10^{-6}} = 2$. The bacterium is 2 micrometers long.

8 Taxi A: \$3/mile. Taxi B: $\frac{36}{9} =$ \$4/mile. Taxi B is more expensive per mile.

9 $(0, -4)$ gives $b = -4$. With $m = \frac{2}{3}$: $y = \frac{2}{3}x - 4$.

10 $-x + 5 = 2x - 1$. Add x: $5 = 3x - 1$. Add 1: $6 = 3x$, so $x = 2$. Then $y = -2 + 5 = 3$.

11 $h(5) = 2(5)^2 = 2(25) = 50$.

12 $A(3) = 2(3) + 15 = 21$. $B(3) = 5(3) + 6 = 21$. They are equal at $x = 3$.

13 $2 - 1 = 1$, $5 - 2 = 3$, $10 - 5 = 5$, $17 - 10 = 7$. The differences change, so the rate of change is not constant — nonlinear.

14 Slope: $\frac{12-3}{4-1} = \frac{9}{3} = 3$. Using $(1,3)$: $3 = 3(1) + b \Rightarrow b = 0$. Equation: $y = 3x$.

15 Filling: water level rises (increasing). Soaking: level stays the same (constant). Draining: level drops (decreasing).

16 Each vertex has its x-coordinate negated while y stays the same: $P(1,1) \to P'(-1,1)$, $Q(4,1) \to Q'(-4,1)$, $R(2,4) \to R'(-2,4)$. This is a reflection over the y-axis.

17 Both rectangles have dimensions 6×10. A rotation maps one onto the other, so they are congruent.

18 Add the translation: $(-6+4,\ 2+(-5)) = (-2,-3)$.

19 Congruent figures are a special case of similar figures with $k = 1$. All congruent figures are similar, but not all similar figures are congruent.

20 An exterior angle and its adjacent interior angle are supplementary: $180 - 140 = 40°$.

21 $7^2 + 24^2 = 49 + 576 = 625 = 25^2$. Since $a^2 + b^2 = c^2$, it is a right triangle.

22 $d = \sqrt{(2-(-1))^2 + (6-2)^2} = \sqrt{9+16} = \sqrt{25} = 5$.

23 $V = \frac{1}{3}\pi(4^2)(9) = \frac{1}{3}\pi(144) = 48\pi\ cm^3$.

24 $V = \frac{1}{3}(35)(6) = \frac{210}{3} = 70\ cm^3$.

25 If there is no clear trend or pattern, there is no association between the variables.

26 $m = \frac{19-7}{9-3} = \frac{12}{6} = 2$.

27 Predicted: $y = 2.5(4) + 1 = 11$. Residual = actual − predicted $= 12 - 11 = 1$.

28 Total who passed $= 40$. Total students $= 60$. Fraction $= \frac{40}{60} = \frac{2}{3}$.

29 First blue: $\frac{6}{10}$. Second blue (without replacement): $\frac{5}{9}$. $P = \frac{6}{10} \times \frac{5}{9} = \frac{30}{90} = \frac{1}{3}$.

30 $\binom{10}{2} = \frac{10\times9}{2} = 45$. Order doesn't matter.

Practice Test 5 — Answer Key

1 Rational | 2 B | 3 A | 4 B | 5 C | 6 11 places to the left | 7 A | 8 D

9 C | 10 $(2.5, 3)$ | 11 19 | 12 A | 13 Nonlinear, each output is x^3. | 14 B | 15 A

16 $(0, -4)$ | 17 C | 18 C | 19 C | 20 B | 21 26 | 22 A | 23 B | 24 B

25 C | 26 B

27 At $x = 5$: $y = 30$. At $x = 30$: $y = -45$. The prediction at $x = 5$ is more reliable because it is within the data range; $x = 3$

28 B | 29 Outcomes: HH, HT, TH, TT. $P(\text{at least one H}) = \frac{3}{4}$. | 30 B

Time to Learn!

Review the explanations below, **especially for the questions you missed.**

Understanding why each answer is correct builds stronger problem-solving skills.

Tip: Circle any questions you got wrong, then read their explanation carefully.

Practice Test 5 — Detailed Explanations

1 $\sqrt{144} = 12$ exactly, because $12^2 = 144$. Since 12 is an integer, $\sqrt{144}$ is rational.

2 The repeating block has 2 digits, so multiply by $10^2 = 100$. Then $100x = 63.6363\ldots$ and $x = 0.6363\ldots$. Subtracting: $100x - x = 63$.

3 $4^2 = 16$ and $5^2 = 25$. Since $16 < 22 < 25$, we have $4 < \sqrt{22} < 5$.

4 Area $= \sqrt{20} \times \sqrt{8} = \sqrt{160}$. $\sqrt{160} \approx 12.65$ cm^2. (Note: $\sqrt{160} = 4\sqrt{10} \approx 12.65$.) Choices A and C confuse addition with multiplication.

5 $81 = 9^2$, so 81 is a perfect square.

6 The exponent is -11, so you move the decimal 11 places to the left from 2.5.

7 Same exponent, so add the coefficients: $5.2 + 3.8 = 9.0$. Answer: 9×10^6.

8 If the relationship passes through $(1, 5)$ and $(3, 15)$, then $k = 5$. But $5 \times 2 = 10 \neq 13$, so $(2, 13)$ does not fit.

9 $m = \frac{8-(-2)}{5-0} = \frac{10}{5} = 2$. The y-intercept is -2: $y = 2x - 2$.

10 Subtract the second from the first: $4y = 12$, so $y = 3$. Then $2x + 3 = 8$, $2x = 5$, $x = 2.5$.

11 $f(5) = 2(5) + 9 = 10 + 9 = 19$.

12 Function P's rate of change: $\frac{7-4}{1-0} = 3$. Function Q's slope is 2. Since $3 > 2$, Function P grows faster.

13 $1^3 = 1$, $2^3 = 8$, $3^3 = 27$, $4^3 = 64$. *The differences* $(7, 19, 37)$ *are not constant, confirming nonlinear.*

14 $y = 25(8) + 60 = 200 + 60 = 260$.

15 *The ball goes up, reaches a peak, then comes down. Gravity causes the rate of change to vary, so the path is curved (nonlinear).*

16 *A* $180°$ *rotation uses* $(x, y) \to (-x, -y)$. *So* $(0, 4) \to (0, -4)$.

17 *In* $\triangle ABC$: $\angle C = 180 - 55 - 80 = 45°$. *Since* $\angle C$ *corresponds to* $\angle F$ *in congruent triangles,* $\angle F = 45°$.

18 *Reflecting over the* y*-axis negates* x *and keeps* y: $(-4, 2) \to (4, 2)$.

19 *All squares have four* $90°$ *angles. Any two squares can be mapped by a dilation since the ratio of their sides is constant. So all squares are similar.*

20 *Corresponding angles are in the same position (e.g., both upper-left) at each intersection of the transversal with the parallel lines.*

21 $c^2 = 10^2 + 24^2 = 100 + 576 = 676$, *so* $c = \sqrt{676} = 26$.

22 $d = \sqrt{(4 - (-2))^2 + (9 - 1)^2} = \sqrt{6^2 + 8^2} = \sqrt{36 + 64} = \sqrt{100} = 10$.

23 $V = \frac{1}{3}(3.14)(64)(24) = \frac{1}{3}(4{,}823.04) \approx 1{,}607.7\ in^3$.

24 $V = \frac{1}{3}Bh = \frac{1}{3}(36)(10) = 120\ cm^3$.

25 *A linear association means the data points roughly follow a straight-line trend.*

26 A good fit means the data points are close to the line, with small distances from each point to the line.

27 $y = -3(5) + 45 = 30$ (interpolation). $y = -3(30) + 45 = -45$ (extrapolation — unreliable and gives a negative value that may not make sense).

28 $0.15 \times 200 = 30$.

29 4 outcomes total. Only TT has no heads. $P(\text{at least one H}) = 1 - P(TT) = 1 - \frac{1}{4} = \frac{3}{4}$.

30 $\binom{6}{3} = \frac{6!}{3! \cdot 3!} = \frac{720}{6 \cdot 6} = 20$.

Practice Test 6 — Answer Key

1 Rational | 2 C | 3 B | 4 B | 5 A | 6 B | 7 A | 8 B | 9 B
10 C | 11 A | 12 A | 13 Linear | 14 B | 15 B | 16 C | 17 5 cm | 18 $(-3, 8)$
19 B | 20 40 | 21 B | 22 $\sqrt{13}$ | 23 B | 24 B | 25 C | 26 C | 27 C | 28 B
29 B | 30 B

Time to Learn!

Review the explanations below, **especially for the questions you missed.**

Understanding why each answer is correct builds stronger problem-solving skills.

Tip: Circle any questions you got wrong, then read their explanation carefully.

Practice Test 6 — Detailed Explanations

1 $\frac{5}{6}$ *is a fraction of two integers with a nonzero denominator, so it is rational. Its decimal is* $0.8\overline{3}$*, which repeats.*

2 *The repeating block* 123 *has* 3 *digits, so you multiply by* $10^3 = 1000$.

3 $10^2 = 100$ *and* $11^2 = 121$. *Since* $100 < 110 < 121$, *we have* $10 < \sqrt{110} < 11$.

4 $\sqrt{a} + \sqrt{b} \neq \sqrt{a+b}$. *You must compute each root separately:* $\sqrt{4} = 2$ *and* $\sqrt{9} = 3$, *so the sum is* 5. $\sqrt{13} \approx 3.6 \neq 5$.

5 *Edge* $= \sqrt[3]{343} = 7$ *in., since* $7^3 = 343$.

6 *Move the decimal* 5 *places left:* $9.03 \times 10^{-5} = 0.0000903$.

7 $4 \times (9.5 \times 10^{12}) = 38 \times 10^{12} = 3.8 \times 10^{13}$ *km.*

8 $k = \frac{y}{x} = \frac{9}{2} = 4.5$.

9 *Substitute* $m = -2$ *and* $b = 7$ *into* $y = mx + b$: $y = -2x + 7$.

10 *The solution to a system is the point* (x, y) *that satisfies both equations, which is where the lines intersect.*

11 $f(10) = 7 - 10 = -3$.

12 *Function A is steeper — it rises faster. Its slope is* 2 *compared to Function B's slope of* 1.

13 *A constant increase of* 6 *means the rate of change is constant, which is the defining property of a linear function.*

14 *Starting height* $b = 12$*. It loses* 2 *inches per hour, so* $m = -2$*. The function is* $y = -2x + 12$*.*

15 *A steeper upward slope means faster increase. The morning section is steeper than the afternoon section, so the temperature rose faster in the morning.*

16 *Reflections are rigid transformations that preserve side lengths, angles, area, and perimeter.*

17 $P = 2l + 2w$*. So* $28 = 2(9) + 2w$*, giving* $28 = 18 + 2w$*,* $2w = 10$*,* $w = 5$ *cm.*

18 *Subtract the translation:* $a = 2 - 5 = -3$*,* $b = 7 - (-1) = 8$*. So* $(a, b) = (-3, 8)$*.*

19 *Similar figures have the same shape, which means all corresponding angles are equal. Side lengths, perimeter, and area may differ.*

20 $90 + x + (x + 10) = 180$*. Simplify:* $2x + 100 = 180$*,* $2x = 80$*,* $x = 40$*.*

21 $c^2 = 1^2 + 1^2 = 2$*, so* $c = \sqrt{2}$*.*

22 $d = \sqrt{2^2 + 3^2} = \sqrt{4 + 9} = \sqrt{13}$*.*

23 $V = \frac{1}{3}\pi r^2 h = \frac{1}{3}\pi(9)(12) = \frac{108\pi}{3} = 36\pi\ m^3$*.*

24 $V = \frac{1}{3}Bh = \frac{1}{3}(30)(9) = 90\ cm^3$*.*

25 *The independent (explanatory) variable goes on the* x*-axis; the dependent (response) variable goes on the* y*-axis.*

26 The y-intercept is the y-value when $x = 0$. The line passes through $(0, 3)$, so $b = 3$.

27 A slope of 0 means no change in y as x changes; the line is horizontal.

28 Relative frequency is the ratio of a frequency to the total, often written as a fraction, decimal, or percentage.

29 Total $= 3 + 5 + 2 = 10$. $P(\text{blue}) = \frac{5}{10} = \frac{1}{2}$.

30 The Fundamental Counting Principle says you multiply the number of outcomes for each stage: $m \times n$.

Practice Test 7 — Answer Key

1 $\sqrt{2}$ (or $\sqrt{3}$)
2 C
3 9 and 10
4 B
5 6 cm
6 B
7 317 times
8 C
9 B
10 B
11 5
12 C
13 C
14 B
15 Decreasing
16 D
17 C
18 B
19 B
20 25
21 B
22 D
23 180π cm^3
24 B
25 No. Association (correlation) does not prove causation.
26 B
27 Predicted height at day 10: $y = 0.4(10) + 2 = 6$ cm. Slope: the plant grows 0.4 cm per day. y-intercept: the plant starts
28 B
29 B
30 C

Time to Learn!

Review the explanations below, ***especially for the questions you missed.***

Understanding why each answer is correct builds stronger problem-solving skills.

Tip: *Circle any questions you got wrong, then read their explanation carefully.*

Practice Test 7 — Detailed Explanations

1 $\sqrt{2} \approx 1.414$ *is between* 1 *and* 2 *and is irrational because* 2 *is not a perfect square.* $\sqrt{3} \approx 1.732$ *also works.*

2 $\frac{36}{100} = 0.36$ *(terminates). The correct conversion uses* $99x = 36$*, giving* $\frac{36}{99} = \frac{4}{11}$*. The student treated* $0.\overline{36}$ *as if it were* 0.36*.*

3 $9^2 = 81$ *and* $10^2 = 100$*. Since* $81 < 83 < 100$*, we have* $9 < \sqrt{83} < 10$*.*

4 $2\sqrt{5} = \sqrt{4} \cdot \sqrt{5} = \sqrt{4 \times 5} = \sqrt{20}$*. Alternatively,* $2\sqrt{5} \approx 4.47$ *while* $\sqrt{10} \approx 3.16$*. They are not equal.*

5 *Edge* $= \sqrt[3]{216} = 6$ *cm, since* $6 \times 6 \times 6 = 216$*.*

6 *Move the decimal* 5 *places left:* $350{,}000 = 3.5 \times 10^5$*. In scientific notation,* a *must satisfy* $1 \leq a < 10$*.*

7 $\frac{1.9 \times 10^{27}}{6 \times 10^{24}} = \frac{1.9}{6} \times 10^3 \approx 0.317 \times 10^3 = 317$*.*

8 *In choice C,* $\frac{9}{3} = 3$ *but* $\frac{15}{6} = 2.5$*. The ratio is not constant, so it is not proportional.*

9 *In* $y = mx + b$*,* $m = 5$ *is the slope and* $b = -3$ *is the* y*-intercept.*

10 $-2x+10=x+1$. *So* $9=3x$, $x=3$. *Then* $y=3+1=4$. *Solution:* $(3,4)$.

11 $f(6)=\frac{6+4}{2}=\frac{10}{2}=5$.

12 $\frac{9-1}{2-0}=\frac{8}{2}=4$. *The rate of change is* 4.

13 $A=s^2$ *has the variable raised to the 2nd power. For a function to be linear, the variable must have an exponent of 1.*

14 *Slope:* $\frac{20-8}{3-0}=\frac{12}{3}=4$. y*-intercept:* 8. *So* $y=4x+8$.

15 *A graph that goes downward means the output (gas) is getting smaller over time.*

16 *A reflection mirrors (flips) a figure across a line of reflection.*

17 *Two figures are congruent if one can be mapped onto the other using translations, reflections, and/or rotations — all rigid transformations.*

18 *Divide the image coordinates by the scale factor:* $(9\div 3,\ -15\div 3)=(3,-5)$.

19 *Any circle can be mapped onto any other circle by a translation (to align centers) followed by a dilation (to match radii).*

20 $(2x+5)+3x+(x+25)=180$. *Combine:* $6x+30=180$, $6x=150$, $x=25$.

21 $a^2=17^2-8^2=289-64=225$, *so* $a=15$. *(Note:* $\sqrt{225}=15$, *so B and D are the same value, but B is simplified.)*

22 $d=\sqrt{(4-1)^2+(5-1)^2}=\sqrt{9+16}=\sqrt{25}=5$.

23 $V = \pi(6^2)(5) = \pi(36)(5) = 180\pi\ cm^3$.

24 $V = \frac{1}{3}(12 \times 5)(8) = \frac{1}{3}(60)(8) = \frac{480}{3} = 160\ ft^3$.

25 *Two variables may be associated due to a third variable or coincidence. Causation requires controlled experiments.*

26 *A straight line of best fit only works for data with a linear pattern. Curved data needs a different model.*

27 *Slope* $= 0.4$ *means* 0.4 *cm of growth per day.* y*-intercept* $= 2$ *means the initial height is* 2 *cm.*

28 *A two-way frequency table displays data for two categorical variables simultaneously, showing how frequently each combination occurs.*

29 $P(A \text{ and } B) = 0.6 \times 0.4 = 0.24$

30 $2^5 = 32$. *Each flip has* 2 *outcomes.*

Practice Test 8 — Answer Key

1 C | 2 $\frac{81}{99} = \frac{9}{11}$ | 3 ≈ 6.7 | 4 D | 5 -5 | 6 C | 7 B | 8 A | 9 B

10 $(4, 5)$ | 11 C | 12 C | 13 $y = x^2$

14 $y = 12x + 50$; $m = 12$ *is the hourly wage,* $b = 50$ *is the base pay.*

15 *Increasing, decreasing, increasing.*

16 $(-a, -b)$ | 17 $55°$ | 18 C | 19 $\frac{3}{2}$ | 20 C | 21 13 *ft* | 22 B | 23 B | 24 6 *in*

25 *Yes. Example: population growth over time may curve upward (exponential), showing a positive but nonlinear as*

26 C	27 25 hours	28 B	29 $0.50 \times 0.20 = 0.10 = 10\%$	30 B

Time to Learn!

*Review the explanations below, **especially for the questions you missed.***

Understanding why each answer is correct builds stronger problem-solving skills.

***Tip:** Circle any questions you got wrong, then read their explanation carefully.*

Practice Test 8 — Detailed Explanations

1 Every integer n can be written as $\frac{n}{1}$, making it rational. Not every square root is irrational (e.g., $\sqrt{4} = 2$).

2 Following the same pattern: $99x = 81$, so $x = \frac{81}{99}$. Simplify by dividing both by 9: $\frac{81}{99} = \frac{9}{11}$.

3 $6.7^2 = 44.89$ and $6.8^2 = 46.24$. Since 45 is very close to 44.89, $\sqrt{45} \approx 6.7$.

4 $\sqrt{3} \times \sqrt{12} = \sqrt{36} = 6$ and $\sqrt{6} \times \sqrt{6} = (\sqrt{6})^2 = 6$. But $\sqrt{3} + \sqrt{12} \approx 1.73 + 3.46 = 5.19 \neq 6$.

5 $(-5)^3 = -125$, so $\sqrt[3]{-125} = -5$.

6 $1 \times 10^2 = 100$ kg. From the chart, the lion's bar reaches close to the 10^2 mark, making its mass closest to 100 kg.

7 $\frac{6.4 \times 10^{10}}{4 \times 10^6} = 1.6 \times 10^4 = 16{,}000$ photos.

8 Machine A: 12 bottles/hour. Machine B: $\frac{50}{5} = 10$ bottles/hour. Machine A is faster.

9 $\frac{6}{3} = 2$ and $\frac{10}{5} = 2$. Similar triangles on the same line always produce the same slope.

10 Subtract: $y = 5$. Then $x + 5 = 9$, so $x = 4$.

11 Find the column where $f(x) = 15$. That column has $x = 4$.

12 Function R: when $x = 0$, $y = 1$. Function S: $b = 1$. Both start at 1, so the initial values are equal.

13 Any equation where x has an exponent other than 1 (such as x^2, x^3, $\frac{1}{x}$, $\sqrt{x}$) is nonlinear.

14 Slope: $62 - 50 = 12$ (hourly rate). y-intercept: 50 (base pay before any hours). Equation: $y = 12x + 50$.

15 Filling: the water level rises. Drinking: the level drops. Refilling: the level rises again.

16 Reflect over x-axis: $(a, b) \to (a, -b)$. Then reflect over y-axis: $(a, -b) \to (-a, -b)$. This is the same as a $180°$ rotation.

17 $\angle R = 180 - 72 - 53 = 55°$. Since $\angle R$ corresponds to $\angle U$, $\angle U = 55°$.

18 The rule $(x, y) \to (-y, x)$ is a $90°$ counterclockwise rotation around the origin.

19 $\frac{9}{6} = \frac{3}{2}$, $\frac{12}{8} = \frac{3}{2}$, $\frac{15}{10} = \frac{3}{2}$. The scale factor is $\frac{3}{2}$.

20 By the Exterior Angle Theorem, y equals the sum of the two non-adjacent interior angles: $y = 55+75 = 130°$.

21 $d = \sqrt{12^2 + 5^2} = \sqrt{144 + 25} = \sqrt{169} = 13$ ft.

22 The distance formula uses the horizontal and vertical distances as legs of a right triangle and applies $a^2+b^2=c^2$.

23 $V=\frac{4}{3}\pi r^3=\frac{4}{3}(3.14)(27)=\frac{4}{3}(84.78)=113.04\ cm^3$.

24 $48=\frac{1}{3}(s^2)(4)$, $48=\frac{4s^2}{3}$, $144=4s^2$, $s^2=36$, $s=6$ *in.*

25 An association can be positive (both increase) and nonlinear (the trend is curved, not straight).

26 The line of best fit does not need to pass through any actual data point. It just needs to be close to most of them.

27 Set $y=0$: $0=-2x+50$, $2x=50$, $x=25$.

28 $\frac{22}{40}=0.55=55\%$.

29 Spins are independent. $P=0.50\times0.20=0.10$.

30 $P(8,3)=8\times7\times6=336$. Order matters.

Practice Test 9 — Answer Key

1 C
2 B
3 A
4 C
5 B
6 B
7 C
8 10 packages
9 B
10 $(2,5)$
11 C
12 A
13 D
14 B
15 C
16 B
17 A
18 B
19 B
20 A
21 C
22 B
23 B
24 B
25 C
26 C
27 B
28 A
29 B
30 $26\times25\times10\times9=58{,}500$

Time to Learn!

*Review the explanations below, **especially for the questions you missed**.*

Understanding why each answer is correct builds stronger problem-solving skills.

***Tip:** Circle any questions you got wrong, then read their explanation carefully.*

Practice Test 9 — Detailed Explanations

1 $1.41421356\ldots$ is the decimal expansion of $\sqrt{2}$, which is non-repeating and non-terminating, indicating an irrational number.

2 Let $x = 0.444\ldots$. Then $10x = 4.444\ldots$. Subtract: $9x = 4$, so $x = \frac{4}{9}$. Note that $\frac{4}{10} = 0.4$ (terminates), which is different from $0.\overline{4}$.

3 $6^2 = 36$ and $7^2 = 49$. Since $36 < 45 < 49$, we have $6 < \sqrt{45} < 7$. The student is correct.

4 $\sqrt{2} \approx 1.414$ and $\sqrt{3} \approx 1.732$, so their sum is approximately 3.146. $\sqrt{10} \approx 3.162$. Since $3.162 > 3.146$, $\sqrt{10}$ is slightly larger.

5 $7^2 = 49$ and $8^2 = 64$. Since $49 < 50 < 64$, we have $7 < \sqrt{50} < 8$.

6 $8 \times 10^5 = 800{,}000$ and $3 \times 10^6 = 3{,}000{,}000$. The higher exponent wins: 3×10^6 is larger.

7 $7 \times 8 = 56$ and $10^{-3} \times 10^{-2} = 10^{-5}$. Then $56 \times 10^{-5} = 5.6 \times 10^{-4}$.

8 Service P: $4 \times 10 = 40$ packages. Service Q: $k = \frac{10}{2} = 5$, so $5 \times 10 = 50$ packages. Difference: $50 - 40 = 10$.

9 A line through the origin has $b = 0$. Only $y = -3x$ has no constant term added.

10 Both tables give $y = 5$ when $x = 2$. The solution is $(2, 5)$.

11 $f(-3) = -(-3) + 8 = 3 + 8 = 11$.

12 The rate of change is the slope. Function A has slope 5 and Function B has slope 2. Since $5 > 2$, Function A has the greater rate of change.

13 $y = \sqrt{x}$ involves a square root of the variable, so it is nonlinear. The others all fit $y = mx + b$.

14 Slope: $\frac{11-5}{2-0} = \frac{6}{2} = 3$. The y-intercept from $(0, 5)$ is $b = 5$. So $y = 3x + 5$.

15 A function can decrease while staying positive (e.g., dropping from 10 to 5). "Decreasing" means the values get smaller, not that they are negative.

16 Translations preserve all side lengths. The longest side is still 8 cm.

17 Both translation and rotation are rigid transformations. The image is congruent to the original regardless of the shape.

18 The $90°$ CCW rule is $(x, y) \to (-y, x)$. So $(-3, 5) \to (-5, -3)$.

19 When $k = 1$, every point stays in the same place. The image is congruent (identical in size and shape) to the original.

20 $\angle C = 180 - 45 - 85 = 50°$. All three angles are less than $90°$, so the triangle is acute.

21 $9^2 + 40^2 = 81 + 1600 = 1681 = 41^2$. The converse of the Pythagorean Theorem confirms it's a right triangle.

22 $d = \sqrt{3^2 + 4^2} = \sqrt{9 + 16} = \sqrt{25} = 5$.

23 $80\pi = \frac{1}{3}\pi(16)h$. Divide both sides by π: $80 = \frac{16h}{3}$. Multiply by 3: $240 = 16h$. Divide: $h = 15$ cm.

24 $V = \frac{1}{3}Bh = \frac{1}{3}(40)(12) = 160$ m^3.

25 The dots go from lower-left to upper-right in a roughly straight pattern. This is a positive linear association.

26 $y = 1.5(4) + 2 = 6 + 2 = 8$.

27 Residual $=$ actual $-$ predicted $= 8 - 6 = 2$.

28 Of 25 who did not study, 10 passed. $\frac{10}{25} = 0.40 = 40\%$.

29 $P = \frac{1}{2} \times \frac{1}{2} \times \frac{1}{2} = \frac{1}{8}$.

30 Letters: 26×25 (no repeat). Digits: 10×9 (no repeat). Total: $26 \times 25 \times 10 \times 9 = 58{,}500$.

Practice Test 10 — Answer Key

1 B
2 D
3 B
4 D
5 C
6 Cell C, Cell A, Cell B
7 3×10^6
8 99
9 C
10 B
11 9
12 B
13 Differences are all 4; the function is linear.
14 B
15 B
16 $(-5, -3)$
17 C
18 $(-5, -2)$
19 A
20 B
21 B
22 B
23 ≈ 226.08 in^3
24 C
25 A
26 Outliers can pull the line away from the overall trend, making it less representative of most data points.
27 B
28 B
29 C
30 B

Time to Learn!

Review the explanations below, ***especially for the questions you missed.***

Understanding why each answer is correct builds stronger problem-solving skills.

Tip: *Circle any questions you got wrong, then read their explanation carefully.*

Practice Test 10 — Detailed Explanations

1 5 is not a perfect square, so $\sqrt{5}$ is irrational and belongs in the Irrational region.

2 In Step 5, the student divided 2 by 10 instead of 9. From $9x = 2$, the correct answer is $x = \frac{2}{9}$.

3 $7^2 = 49$ and $8^2 = 64$. Since $49 < 50 < 64$, we have $7 < \sqrt{50} < 8$.

4 $\pi \approx 3.1416$, $\sqrt{10} \approx 3.1623$, and 3.2 is exact. From least to greatest: $3.1416 < 3.1623 < 3.2$.

5 $\sqrt{2}$ cannot be expressed as a fraction of two integers, so it is irrational.

6 Cell A: 8×10^{-3} mm. Cell B: 5×10^{-2} mm. Cell C: 3×10^{-4} mm. From smallest to largest: $3 \times 10^{-4} < 8 \times 10^{-3} < 5 \times 10^{-2}$, so Cell C, Cell A, Cell B.

7 $\frac{3.6}{1.2} = 3$ and $10^{10-4} = 10^6$. Answer: 3×10^6.

8 $k = \frac{54}{6} = 9$. When $x = 11$: $y = 9 \times 11 = 99$.

9 Through the origin, $b = 0$. Slope $= \frac{12}{3} = 4$. Equation: $y = 4x$.

10 Add: $2x = 14$, so $x = 7$. Then $y = 10 - 7 = 3$. Solution: $(7, 3)$.

11 From the graph, $f(1) = 3$ and $f(4) = 6$. So $f(1) + f(4) = 3 + 6 = 9$.

12 Set $2x + 100 = 8x + 10$. Subtract $2x$: $100 = 6x + 10$. Subtract 10: $90 = 6x$. Divide: $x = 15$.

13 $1 - (-3) = 4$, $5 - 1 = 4$, $9 - 5 = 4$, $13 - 9 = 4$. The constant difference of 4 means the rate of change is constant, so the function is linear.

14 Starting amount $b = 120$. Spending \$15/day means losing money, so $m = -15$. The function is $y = -15x + 120$.

15 A steeper distance-time graph means more distance covered per unit of time — that means greater speed.

16 Reflecting over the y-axis flips the sign of the x-coordinate: $(5, -3) \to (-5, -3)$.

17 Two circles with the same radius are always congruent — one can be translated to overlap the other.

18 $90°$ CCW: $(x, y) \to (-y, x)$. So $(-2, 5) \to (-5, -2)$.

19 Perimeter scales by the same factor: $36 \times \frac{1}{3} = 12$ cm.

20 Co-interior angles are supplementary: $180 - 135 = 45°$.

21 $c^2 = 6^2 + 8^2 = 36 + 64 = 100$, so $c = 10$.

22 The x-coordinates are the same, so the distance is the vertical difference: $|7 - (-5)| = 12$.

23 $V = 3.14 \times 9 \times 8 = 226.08\ in^3$.

24 Prism volume $= 6 \times 6 \times 10 = 360\ in^3$. Pyramid volume $= \frac{360}{3} = 120\ in^3$.

25 Clusters are groups of points that bunch together, while outliers are isolated far from the trend. Both can appear in the same scatter plot.

26 An outlier is an unusual point. Including it would distort the line and give a worse fit for the majority of the data.

27 The y-intercept ($b = 10$) is the predicted value when $x = 0$, meaning the predicted score with no study time.

28 Boys: 6 summer vs. 3 winter. Girls: 4 summer vs. 5 winter. Boys lean toward summer, girls lean toward winter.

29 $P(6) = \frac{1}{6}$. Independent: $P = \frac{1}{6} \times \frac{1}{6} = \frac{1}{36}$.

30 $\binom{5}{2} = \frac{5!}{2! \cdot 3!} = \frac{120}{2 \cdot 6} = 10$.

Well done checking your answers!

Keep practicing to strengthen your skills.

Author's Final Note

I hope you enjoyed this book as much as I enjoyed writing it. Whether you are a student working through the mater a parent supporting your child's learning, or a teacher guiding your class, I have tried to make this book as clear an engaging as possible. I hope I have succeeded. If you have any suggestions for improvement, please let me know. I would love to hear from you.

The accuracy of calculations is very important to me. We have done our best, but I also expect that I have made sor minor errors. Constant improvement is the name of the game. If you find any errors, please let me know. I will fix them in the next edition.

***For students:** Your learning journey does not end here. I have written a series of books to help you learn math. Mal sure you browse through them. I especially recommend workbooks and practice tests to help you prepare for your exams.*

***For parents:** Thank you for investing in your child's education. I encourage you to explore the companion resources available online to help support your child outside the classroom.*

***For teachers:** Thank you for the invaluable work you do every day. I hope this book serves as a useful resource in yo classroom. Feel free to reach out if you have suggestions or would like to discuss how best to use this book with your students.*

I also enjoy reading your reviews. If you have a moment, please leave a review on where you found this book. It will help others find this book. If you have any questions or comments, please feel free to contact me at drNazari@ViewMath.com.

And one last thing: Remember to use online resources for additional help. I recommend using the resources on `https://ViewMath.com` You can find video lessons, practice problems, and more. You can also use the online companion for this book to track your progress and access additional resources.

Wishing all students the best in their studies, parents every success in supporting their children, and teachers continued inspiration in their classrooms!

Dr. A. Nazari